ORIGIN AND RELATIONSHIPS OF THE CALIFORNIA FLORA

ORIGIN AND RELATIONSHIPS OF THE CALIFORNIA FLORA

BY PETER H. RAVEN AND DANIEL I. AXELROD

 CALIFORNIA NATIVE PLANT SOCIETY

Copyright ©1978, 1995 by the Regents of the University of California. Originally published in the University of California Publications in Botany, Volume 72. University of California Press.

Cover design: Beth Hansen

Printed in the United States of America
ISBN: 0-943460-27-1

Dedicated to the Memory of
WILLIS LINN JEPSON
1867-1946

Through a lifetime of perceptive studies he gave us the first clear understanding of the flora of California.

Looking eastward from the summit of Pacheco Pass one shining morning, I found before me a landscape . . . that after all my wanderings still appears as the most beautiful I have ever beheld. At my feet lay the Great Central Valley of California, level and flowery, like a lake of pure sunshine, forty or fifty miles wide, five hundred miles long, one rich furred garden of yellow compositae. And from the eastern boundary of this vast golden flower-bed rose the mighty Sierra, miles in height, and so gloriously colored and so radiant, it seemed not clothed with light, but wholly composed of it, like the wall of some celestial city. Along the top and extending a good way down was a rich pearl-gray belt of snow; below it a belt of blue and dark purple, marking the extension of the forests; and stretching along the base of the range a broad belt of rose-purple; all these colors, from the blue sky to the yellow valley, smoothly blending as they do in a rainbow, making a wall of light ineffably fine. . . .

John Muir, April, 1868 (Muir, 1912)

CONTENTS

ORIGIN AND RELATIONSHIPS
OF THE CALIFORNIA FLORA

by

Peter H. Raven[1] and Daniel I. Axelrod[2]

INTRODUCTION

The flora of California, one of the most famous in the world, is a mixture of northern, temperate elements and xeric, southern elements, and is characterized by a very high degree of endemism. That it was a mixture of two distinct floristic elements was recognized clearly by Abrams (1925), elaborated by Campbell and Wiggins (1947), and restated on the basis of a detailed knowledge of the fossil record by Axelrod (1958). In the present monograph, we shall elaborate this theme, and also consider the reasons for the large number of species and the high proportion of endemism in the flora. Part of the explanation is found in the equable climate that has prevailed in California throughout most of the Tertiary, as pointed out by Abrams (1926), among others. We shall emphasize that another important factor has been the Late Pliocene to Recent elevation of the Sierra Nevada and other ranges, together with the concomitant development of a cold off-shore current which ultimately resulted in the development of a mediterranean, summer-dry climate during the past million years, as pointed out by Axelrod (1966). The endemics of California are a mixture of relicts and newly produced species, as outlined by Stebbins and Major (1965), and it is the latter that have contributed most to the size of the flora and to the high proportion of endemism in it. Earlier, highly condensed versions of this monograph were published by Raven (1977) and Axelrod (1977) in a volume on the terrestrial vegetation of the State.

NUMBERS OF GENERA AND SPECIES

California is very rich in species of native vascular plants, with 5046, but not especially rich in native genera, with 878 (table 1). It shares this richness in species with other areas of the world that also have a mediterranean climate. The California flora appears even more impressive when the number of genera and species in the California Floristic Province (CFP), which consists chiefly of that part of the State west of the Sierra Nevada-Cascade axis, and excluding the deserts, is considered (Howell, 1957). The California Floristic Province (fig. 1), with an area of about 285,000 km^2 in the State of California, has 748 genera and 4119 species of native vascular plants in its California portion. There are about 130 genera and 927 species in the State that do not occur within the California Floristic Province.

Portions of Oregon and Baja California are also assigned to the California Floristic Province on the basis of their common flora and climate (Jepson, 1925, p. 3; Howell, 1957). In Oregon, the boundary runs from Coos Bay southeast to the divide between the Umpqua and Rogue rivers and thence up the arm of the Rogue River Valley in which Ashland is situated to the California line. This is an area of approximately 25,000 km^2. An analysis of

[1] Missouri Botanical Garden, 2345 Tower Grove Ave., St. Louis, Mo., 63110.
[2] Department of Botany, University of California, Davis, Calif., 95616.

TABLE 1

Native Genera and Species of Vascular Plants in Various Regions

	Area (1000's of km^2)	Genera	Endemic %	Species	Endemic %
California[1]	411	878	26 (3.0)	5046	1517 (30.1)
California Floristic Province (CFP)[2]	324	795	50 (6.3)	4452	2125 (47.7)
Alaska[3]	1479	355	0 (0)	1366	80 (5.9)
Barro Colorado Island[4]	0.016	652	0 (0)	1261	?
British Isles[5]	308	545	0 (0)	1443	17 (1.2)
Cape Peninsula, South Africa[6]	0.47	533	1 (0.2)	2256	157 (7.0)
Carolinas[7]	217	819	1 (0.1)	2995	23 (0.8)
Galápagos Islands[8]	7.9	250	7 (2.8)	701	175 (25.0)
Gray's Manual Area[9]	3238	849	6 (0.7)	4425	599 (13.5)
Guatemala[10]	109	1799	?	7817	?
Hawaii[11]	16.6	253	31 (12.3)	1897	1751 (92.3)
Japan[12]	377	1098	17 (1.5)	4022	1371 (34.1)
New Zealand[13]	268	393	39 (9.9)	1996	1618 (81.1)
Sonoran Desert[14]	310	746	20 (2.7)	2441	650 (26.6)
Texas[15]	751	1075	7 (0.7)	4196	379 (9.0)
World	–	13,600	–	255,000	–

[1] Smith and Noldecke, 1960; Howell, 1972. Figures derived from Munz and Keck, 1959, and Munz, 1968, modified to agree with treatments of Loasaceae (Thompson and Roberts, 1974) and Cactaceae (Benson, 1969), as in Munz, 1974, and with an original, unpublished treatment of Onagraceae.

[2] Including 90 additional species and 40 additional endemics in S.W. Oregon and 227 additional species and 107 additional endemics in N.W. Baja California, the latter totals kindly furnished by Reid Moran.

[3] Hultén, 1968.

[4] Croat, 1977.

[5] Clapham, Tutin, and Warburg (1962) give 2375 species less 673 adventives, but we have adopted the figures given by Dandy (1958); he also lists 643 microspecies of *Alchemilla* (1), *Hieracium* (223), *Limonium* (3), *Rubus* (389), *Sorbus* (15), and *Taraxacum* (25), the number of microspecies in each genus being given in parentheses. (D. H. Valentine, personal communication).

[6] Adamson and Salter, 1950; estimates of endemics by A. V. Hall.

[7] Radford, Ahles, and Bell, 1968; calculated by David Boufford.

[8] Wiggins and Porter, 1971; Johnson and Raven, 1973; Porter, 1976 and personal communication.

[9] Fernald, 1950; numbered microspecies in *Rubus* and *Crataegus* included in species total. The 599 endemics include 164 species of *Rubus* and 67 species of *Crataegus*.

[10] Estimates for ferns by R. G. Stolze, the remainder from Standley and Williams, 1946-1976; all spontaneously reproducing plants included.

[11] Fosberg, 1948; Hillebrand, 1888. Totals include subspecies and varieties.

[12] Ohwi, 1965; 4263 species less 241 adventives. No distinction made for microspecies.

[13] Godley, 1975.

[14] Shreve and Wiggins, 1964. Many species not actually said to occur in the Sonoran Desert were subtracted from the totals used here.

[15] Correll and Johnston, 1970; 141 genera and 643 species of clearly introduced plants subtracted from their totals, but numbers for native genera and species probably still too high.

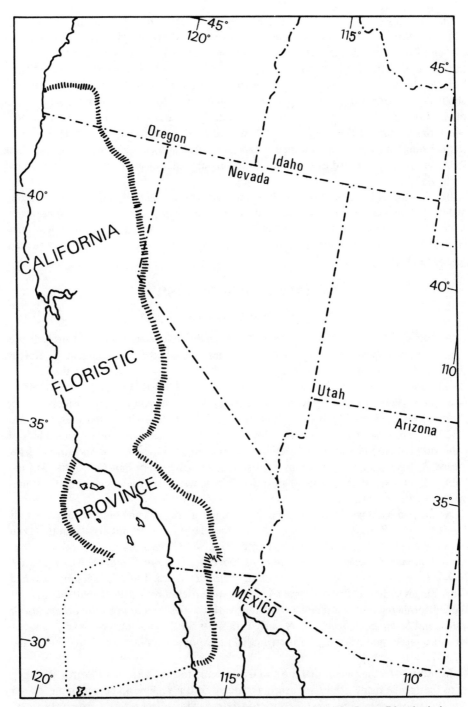

Fig. 1. The California Floristic Province (CFP), extending from the Rogue River basin in Oregon, southward west of the deserts to El Rosario in northwestern Baja California, and including Guadalupe Island.

the data in Peck (1961) shows that there are approximately 90 species in this area not found in California. In Baja California, the California Floristic Province includes the chaparral and forest belts of the Sierra Juárez and the Sierra San Pedro Mártir, but not their eastern desert flanks, extends south along the coast to near El Rosario (Howell, 1957), and includes Guadalupe Island also. Approximately 227 species in this area, which occupies about 14,000 km², do not occur within the borders of California (R. Moran, personal communication). Excluded are Cedros Island (Moran, 1972) and various mainland peaks of which the Volcán las Tres Vírgenes (ca. 1995 m elevation) at about 28°35 N lat. is the southernmost station for several species typical of the California Floristic Province. These localities are best included in a general phytogeographic classification with their surrounding lowlands.

The California Floristic Province thus comprises about 324,000 km², and has approximately 794 genera and 4452 species of native vascular plants. This is more than in the entire central and northeastern United States and adjacent Canada, a region some ten times larger! This region clearly contains the largest and most diverse assemblage of native plant species in all of temperate and northern North America.

SURVEY OF ENDEMISM

Endemic Genera

For the state of California as a whole, 26 of the 878 native genera (3.0%) are endemic. Three of these—*Gilmania* (Polygonaceae: 1 species), *Dedeckera* (Polygonaceae: 1; Reveal and Howell, 1976), and *Swallenia* (*Ectosperma*, Poaceae: 1)—are endemic to the Inyo County region beyond the boundaries of the California Floristic Province. Three more—*Goodmania* (Polygonaceae: 1; Reveal and Ertter, 1977), *Mucronea* (Polygonaceae: 2), and *Parishella* (Lobeliaceae: 1)—are confined to the state of California but range onto the Mohave Desert or reach the mountains of Inyo County; they are therefore not endemic to the California Floristic Province. *Malacothamnus* (Malvaceae: 11) and *Tropidocarpum* (Brassicaceae: 2) have similar distributions but also reach northwestern Baja California. Within the California Floristic Province, there are about 795 genera of native vascular plants, of which 50 (6.3%) are endemic (table 2).

As stressed by Howell (1957), the impressive nature of the endemic genera in the California Floristic Province stems not only from their large number, but also from the nature of the genera involved. Many—such as *Carpenteria, Hesperelaea, Jepsonia* (Ornduff, 1969c), *Lyonothamnus, Odontostomum, Romneya, Sarcodes, Sequoia, Sequoiadendron,* and *Umbellularia*—are outstanding relicts without any close relatives. These genera doubtless had wider ranges earlier in Tertiary time and are now restricted to relatively mild climates in the California region, an inference consistent with the fossil record of *Lyonothamnus, Sequoia,* and *Sequoiadendron.* Others, such as most of the 22 endemic genera of Asteraceae, are more recent and probably have originated in response to spreading aridity during and after Neogene time.

In addition to the genera that are strictly endemic to the California Floristic Province or to California, many others are nearly restricted to this region. Some, including *Allophyllum* (Polemoniaceae: 5 species), *Fremontodendron* (Sterculiaceae: 2), *Lessingia* (Asteraceae-Astereae: 12), *Machaerocarpus* (Alismataceae: 1), *Muilla* (Liliaceae: 3), *Pholistoma* (Hydrophyllaceae: 3), and *Turricula* (Hydrophyllaceae: 1), range beyond the borders of

TABLE 2
Genera Endemic to the California Floristic Province[1]

Acanthomintha (Menthaceae: 3)–S. Coast Ranges, Calif. to N.W. Baja Calif.
Achyrachaena (Asteraceae, Madiinae: 1)–Douglas Co., Ore. to S. Calif.
Adenothamnus (Asteraceae, Madiinae: 1)–N.W. Baja Calif.
Baeriopsis (Asteraceae, Helenieae: 1)–Guadalupe I., Baja Calif.
*Benitoa (Asteraceae, Astereae: 1)–S. Coast Ranges, Calif.
Bensoniella (Saxifragaceae: 1)–S.W. Ore. to N.W. Calif.
Bergerocactus (Cactaceae: 1)–S. Calif. to N.W. Baja Calif.
*Blepharizonia (Asteraceae, Madiinae: 1)–S. Coast Ranges, Calif.
Bloomeria (Liliaceae: 2)–Calif. to N. Baja Calif.
Calycadenia (Asteraceae, Madiinae: 12)–N. Umpqua Valley, Ore. to N. Baja Calif.
*Carpenteria (Saxifragaceae: 1)–Cent. Sierra Nevada, Calif.
Chamaebatia (Rosaceae: 2)–Calif. to N. Baja Calif.
Chlorogalum (Liliaceae: 5)–Coos Co., Ore., to N.W. Baja Calif.
Corethrogyne (Asteraceae, Astereae: 3)–Coos Co., Ore. to N.W. Baja Calif.
Dendromecon (Papaveraceae: 2)–Calif. to N.W. Baja Calif.
*Draperia (Hydrophyllaceae: 1)–N. Calif.
*Eastwoodia (Asteraceae, Astereae: 1)–S. Coast Ranges, Calif.
Hesperelaea (Oleaceae: 1)–Guadalupe I., Baja Calif.
Hesperomecon (Papaveraceae: 1)–S.W. Ore. to N.W. Baja Calif.
Heterodraba (Brassicaceae: 1)–S.W. Ore. to N.W. Baja Calif.
Heterogaura (Onagraceae: 1)–S.W. Oregon to Calif.
*Hollisteria (Polygonaceae: 1)–Calif.
*Holocarpha (Asteraceae, Madiinae: 4)–Calif.
*Holozonia (Asteraceae, Madiinae: 1)–Calif.
Jepsonia (Saxifragaceae: 3)–Calif. to N.W. Baja Calif.; Ornduff, 1969c.
Lembertia (Asteraceae, Helenieae: 1; *Eatonella* in part)–Calif.
Lemmonia (Hydrophyllaceae: 1)–Calif. and adjacent W. Nev. to N. Baja Calif.
*Lyonothamnus (Rosaceae: 1)–Islands of S. Calif.
*Monolopia (Asteraceae, Helenieae: 4)–Calif.
*Neostapfia (Poaceae: 1)–Great Valley, Calif.
*Odontostomum (Liliaceae: 1)–Calif.
Ophiocephalus (Scrophulariaceae: 1)–Sierra San Pedro Mártir, Baja Calif.
*Oreonana (Apiaceae: 2)–Calif.
Ornithostaphylos (Ericaceae: 1)–S. San Diego Co., Calif. to N. Baja Calif.
*Orochaenactis (Asteraceae, Helenieae: 1)–S. Sierra Nevada, Calif.
*Parvisedum (Crassulaceae: 4)–Calif.
Pentachaeta (Asteraceae, Astereae: 6)–Calif. to N. Baja Calif.; van Horn, 1973; Nelson and van Horn, 1976
*Phalacroseris (Asteraceae, Lactuceae: 1)–Sierra Nevada, Calif.
Pickeringia (Fabaceae: 1)–Calif. to N.W. Baja Calif.
Pogogyne (Menthaceae: 5)–S.W. Ore. to N.W. Baja Calif., including Guadalupe I.
*Pseudobahia (Asteraceae, Helenieae: 3)–Calif.
Romneya (Papaveraceae: 2)–S. Calif. to N.W. Baja Calif.
Sarcodes (Ericaceae: 1)–S. Ore. to N. Baja Calif.
Sequoia (Taxodiaceae: 1)–S.W. Oregon to coastal central Calif.
*Sequoiadendron (Taxodiaceae: 1)–Sierra Nevada, Calif.
Stylomecon (Papaveraceae: 1)–Calif. to N.W. Baja Calif.
*Tracyina (Asteraceae, Astereae: 1)–N.W. Calif.
Umbellularia (Lauraceae: 1)–S. Douglas Co., Ore., to S. Calif.
Venegasia (Asteraceae, Helenieae: 1)–S. Calif. to N.W. Baja Calif.
*Whitneya (Asteraceae, Senecioneae: 1)–Calif.

[1] *Munzothamnus* has been reduced to the synonymy of *Stephanomeria* by Tomb (1974). *Congdonia*, described from the east slope of the Sierra Nevada in Mono Co., is probably not distinct from *Sedum* and may have been mislabeled as to locality (Moran, 1950b; Clausen, 1975, p. 68, 186). Genera endemic to the state of California are marked with an asterisk.

the Floristic Province onto the deserts or Great Basin or occur in the woodland and chaparral of Arizona. A number, including *Adenostoma* (Rosaceae: 2), *Aphanisma* (Chenopodiaceae: 1), *Apiastrum* (Apiaceae: 1), *Cneoridium* (Rutaceae: 1), *Heteromeles* (Rosaceae: 1),
Malosma (Anacardiaceae: 1; Young, 1974), and *Xylococcus* (Ericaceae: 1) range southward
to central Baja California, with *Heteromeles* and *Malosma* even in the mountains of the
Cape Region. Others, such as *Darlingtonia* (Sarraceniaceae: 1), *Kalmiopsis* (Ericaceae: 1),
and *Peltiphyllum* (Saxifragaceae: 1), also occur beyond the limits of the California Floristic Province in the Cascades and Coast Ranges of Oregon. The distinctive grass genus *Orcuttia* (7 species) would be endemic, as is its closest relative, the monotypic *Neostapfia*,
were it not for a species of *Orcuttia* which is confined to the Magdalena Plain of southern
Baja California.

The North American species of subtribe Madiinae of Asteraceae, tribe Heliantheae, with
12 genera and 97 species in western North America, all occur within the California Floristic Province and are endemic to it except for a few more widely ranging species of *Madia,
Layia,* and *Raillardella,* one of *Lagophylla* and the single species of *Blepharipappus,* all of
which occur outside the State but also within the boundaries of the California Floristic
Province. In addition, one species of *Hemizonia*, a genus with 30 species endemic in the
California Floristic Province, is restricted to the low, desert San Benito Islands off of Cedros Island, where the genus does not occur. The subtribe also includes three genera and
about a dozen species of the Hawaiian Islands (Carlquist, 1959). A similar pattern is found
in the genus *Hesperocnide* (Urticaceae: 2), with one species in California, a second in Hawaii.

Several genera are characteristic of the California Floristic Province, and have a large
majority of their species there, but also include one or more widely ranging species and are
therefore not endemic; examples are *Arctostaphylos* (Ericaceae: 50, all of which occur in
the California Floristic Province although a few range widely outside of this region, P. V.
Wells, personal communication), *Ceanothus* (Rhamnaceae: 53, 43 in California), *Clarkia*
(Onagraceae: 41, 2 outside of California), *Collinsia* (Scrophulariaceae: 17, 3 outside of
California), *Hemizonia* (mentioned above: 30 of 31 species endemic in California Floristic
Province), and *Hesperolinon* (Linaceae: 12, all in California; Sharsmith, 1961), as well as
the small family Limnanthaceae, with 10 species, only one of which does not occur in the
California Floristic Province. *Blennosperma* (Asteraceae, Senecioneae: 3; Ornduff, 1963,
1964; Skvarla and Turner, 1966; Ornduff et al., 1973) would be endemic to the California
Floristic Province except for a single species in Chile. This pattern is mirrored in *Amblyopappus* (Asteraceae-Helenieae: 1), *Clarkia, Lastarriaea* (Polygonaceae: 1), *Microsteris* (Polemoniaceae: 1), and certain other genera discussed on pp. 40-41 and listed in table 10
(Raven, 1963); they range outside of the California Floristic Province in North America also.

Endemic Species

The percentage of endemism in the California Floristic Province is 47.7%—about 2125
of the 4452 native species. Such a high percentage is very unusual for a continental area,
although it may be equalled or exceeded in the other parts of the world with a mediterranean climate. Certain aspects of this endemic group will now be discussed in relation to
the statistics presented in table 3.

First, California has a relatively high number of species per genus, a concept reviewed
by Simberloff (1970). The region is intermediate in this respect between the other conti-

TABLE 3

Percentage of Species of Monocotyledons, Asteraceae, Annuals,
and Percentage of Species in Ten Largest Genera for Selected Regions,
Using Same Sources as Indicated in Table 1.

	Species/ Genus	Ten Largest Genera	Mono- cotyledons	Asteraceae	Annuals
California	5.7	16.1%	18.1%	12.2%	28.6%
California Floristic Province	5.6	15.2%	19.2%	13.6%	27.4%
Alaska	3.8	26.0%	28.6%	10.0%	2.1%
Barro Colorado Island	1.9	12.1%	27.4%	2.9%	< 3.9%
British Isles	2.65	18.2%	25.0%	8.7%	21.6%
Cape Peninsula	4.2	17.5%	34.6%	11.5%	9.6%
Carolinas	3.5	14.5%	23.6%	10.4%	3.8%
Galápagos Islands	2.8	14.6%	17.0%	6.3%	19.5%
Gray's Manual Area	5.2	21.8%	28.2%	12.7%	8.7%
Guatemala	4.3	?	22.3%	7.7%	?
Hawaii	7.5	42.1%	.8.5%	11.4%	0.04%
Japan	3.7	14.6%	28.0%	8.5%	7.3%
New Zealand	7.4	26.3%	27.3%	12.5%	6.0%
Sonoran Desert	3.3	12.8%	12.1%	15.0%	21.4%
Texas	3.9	10.2%	24.4%	13.4%	20.4%
World	18.7	~6%	25.4%	7.8%	13.0%

nental areas enumerated in table 3 and islands or island groups such as Hawaii and New Zealand, which have far more species per genus than are typical of mainland areas of comparable size. This suggests that bursts of speciation in certain plant genera have made important contributions to the endemism and to the high total number of species in California. On the other hand, the second column of table 3—giving the percentages of the total native flora included in the ten largest genera—indicates that California is not exceptional in this respect. The clusters of species mentioned above are scattered among many more than ten genera, unlike the situation on some oceanic islands.

Second, California has a numerical representation of monocotyledons comparable to that of most other areas enumerated in the table. The lower proportion of monocots compared with many other temperate areas therefore can be attributed to an unusually high proportion of dicots in California, not to lower success of monocots per se in this region. Noteworthy in this column, however, are the especially low numbers of monocots in Hawaii, the Galápagos Islands, and the Sonoran Desert, and the notably high number of monocots which make a very significant contribution to the richness of the flora in the Cape Peninsula of South Africa. In the California Floristic Province, only about 31% of the monocots, but more than 53% of the much more numerous dicots, are endemic. This indicates an even greater contribution of dicots to the endemism of the region than their numbers would suggest.

Third, even though Asteraceae comprise the largest family of flowering plants in California, the proportion of this family in the flora is not exceptional, despite frequent statements to the contrary. The family comprises an eighth of the total number of vascular plants in many regions, including California. Of the 696 species of Asteraceae in California,

217 are endemic (Smith and Noldecke, 1960)—31.2% as compared with 30.2% endemism for the flora of the State as a whole and 31.0% for the dicots alone. Several other large families are more impressive from the standpoint of endemism in the State, among them Malvaceae (53.3%), Menthaceae (49.5%), Rhamnaceae (47.5%), and Scrophulariaceae (41.1%). For the California Floristic Province, about 320 of a total of about 620 Asteraceae are endemic (51.6%), as compared with about 53% for the dicots as a whole.

Fourth, the last column of table 3 shows that the flora of California has an unusually high proportion of annuals—28.6% for the State, 27.4% for the Floristic Province. This is similar to the 30% annuals reported by Orshan (1953) for the Mediterranean vegetation of Palestine, but well above the 13% average reported by Raunkiaer (1934) for the flora of the world and also the 21.3% for North America north of Mexico (3155 of 14,500 species; Shetler and Meadow, 1972). Higher percentages have been reported by Raunkiaer (1934) for Palestine as a whole (48.3%) and for the Libyan Desert (42%), and by Orshan (1953) for the Irano-Turanian (47%) and Saharo-Sindian (60%) vegetation of Palestine. For Egypt, some 960 (47.5%) of the approximately 2022 apparently indigenous species are annual (Täckholm, 1974). This figure contrasts strikingly with that for the Sonoran Desert, where only about 522 of 2441 species (21.4%) are annual. Thus it appears that the proportion of annual vascular plants is indeed considerably higher in the dry regions of the Old World, extending from the Mediterranean Basin to Mongolia, than anywhere in the New World, especially in view of the 21.4% annual species indicated for the Sonoran Desert in table 3. This may be related to the much larger arid area in the Old World, and to its longer history of continentality. Nonetheless, the proportion of annual species in California is high, and contributes greatly both to the high total number of species and to the endemism.

Of the approximately 1098 species of annual dicots in the California Floristic Province, about 695 (63.3%) are endemic. Of the 2440 biennial or perennial species, about 1095 (44.9%) are endemic. Annual monocots are much less common, not only in California, but in general, for reasons that are unknown to us. In fact, in California there are only about 85 native species of annual monocots: 9 of *Juncus* (including 5 endemics), 13 Cyperaceae (none endemic), and 63 Poaceae (including the following endemics: *Neostapfia*, with 1 species; 6 of the 7 species of *Orcuttia*; 1 species each of *Phalaris* and of *Puccinellia*; and 2 of *Agrostis*). Sixteen of the 85 species of the annual monocots of California (18.8%) are endemic, as contrasted with 21% endemism for the monocots of the state in general, or 31% for the California Floristic Province. The evolution of annual species of dicots, but not monocots, has clearly been a significant feature of the evolution of the flora of California. On the other hand, the even more impressive representation of annual species of both groups in the Old World probably underlies the great success of many annual species of Old World origin as weeds throughout the world.

One area in California that has a high proportion of annual species is the Mohave Desert. Excluding such ranges as the Clark, Kingston, and New York Mountains of the eastern Mohave Desert, associated by Stebbins and Major (1965) with the Inyo subdivision of the Great Basin, the Mohave Desert comprises about 45,000 km². Judged from the distributional data given by Munz (1974), there are about 757 species in this area, of which 334 (44.1%) are annuals. This high percentage comes about, however, not because the Mohave Desert is an important center of endemism or evolution of annual plants, but because most of the woody plants, herbaceous perennials, and stem succulents of the Sonoran Desert are unable to live there because of unfavorable winter temperatures and summer drought

(Shreve and Wiggins, 1964). Many genera of annual plants characteristic of cismontane California, such as *Amsinckia, Collinsia, Lasthenia, Layia, Lupinus, Mimulus, Montia,* and *Trifolium,* do range out onto the desert in regions where the winter rainfall is adequate for them. There are about 22 endemic species in the Mohave Desert as defined strictly, or 3.3% of the total flora, and about 90 species of monocots (11.9%), a proportion of monocots similar to that of the Sonoran Desert (12.1%), but very low on a world basis.

CISMONTANE CALIFORNIA:
THE CALIFORNIA FLORISTIC PROVINCE

The California Floristic Province has a rich and remarkable flora of native vascular plants for three principal reasons. *First,* many Tertiary relicts have survived here owing to its sheltered, equable climate. It was uplift of the Sierra Nevada, Peninsular Ranges and Coast Ranges during the Pliocene and Quaternary that provided protection from the more extreme continental climates that developed over the interior following the Pliocene. *Second*, environmental diversity is great because the region is one of high relief, reaching from the margins of subtropical desert to well above timberline. The area stretches from a perhumid temperate rainforest climate in the northwest corner to semiarid and arid hot interior valleys in the south, and the variety of rock type and hence edaphic conditions is notably complex. These diverse conditions provide innumerable local environments that can and do support numerous taxa. *Third,* outbursts of speciation have occurred in certain woody and herbaceous, especially annual dicot, genera, and notably since the Middle Pliocene. This activity was stimulated chiefly by recurrent climatic fluctuations (cool-moist, warm-dry) in a region where an increasing diversity of topographic, climatic, and edaphic conditions developed concurrently (Axelrod, 1966b).

To provide a general background for understanding these dynamic interrelations, we recall briefly some of the salient features of the history of the flora. More detailed analyses are presented elsewhere (Axelrod, 1966a, 1966b, 1967a, 1967b, 1968, 1973, 1975, 1976a, 1976b, 1977).

General Setting

Early in the Tertiary, the region differed greatly in topography, vegetation and climate from that of the present. The high mountains that now dominate the landscape were not yet in existence. A tropical sea lapped against low hills at the site of the present Sierra Nevada and Klamath Mountain region, and also covered most of the Coast Ranges. The portion of California west of the San Andreas was then situated far to the south. The Cape Region of Baja California was nestled against the coast of Nayarit near Cabo Corrientes; the east coast of Baja California lay against the present coast of Sonora/Sinaloa. The proto-Gulf of California did not appear until the Late Miocene when plate movements commenced. Southern California west of the San Andreas was then situated in the present area of northern Baja California. It was covered with tropical seas into the Late Miocene which flooded much of the area of the present Transverse Ranges and lapped against the front of the low San Gabriel Mountains, thence swinging southwest in Riverside County to the present coastline, and then south along the coast of Baja California. In the Miocene and later, the entire area continued to shift northward in a series of complex movements distributed among several major faults of the San Andreas system (Karig and Jensky, 1972; Moore, 1973; Gastil and Jensky, 1973).

Fossil floras of Eocene and Early Oligocene age from Washington southward show that the coastal region was covered by dense rainforest. It lived under a warm temperate, perhumid climate in the north, grading into a subtropical savanna that thrived under monsoonal conditions in central and southern California. Fossil floras from this region are composed of tree ferns, cycads, and numerous large-leaved evergreen dicots of tropical and subtropical families chiefly, including Anacardiaceae, Arecaceae, Burseraceae, Fabaceae-Caesalpinioideae, Lauraceae, Meliaceae, Moraceae, Simaroubaceae, and Sapindaceae. A few taxa of temperate requirements are present and they increase inland, as shown by the appearance of *Alnus, Carya, Cladastris, Metasequoia, Platanus, Quercus, Zelkova* and others in eastern California, Oregon and Washington which contributed to a rich evergreen-deciduous hardwood forest farther inland.

There was a rapid change in forest composition at the edge of the cordilleran region, stretching from central British Columbia southward through Idaho and into central Nevada, and down the Rocky Mountain axis through central Colorado and New Mexico. The region was one of discontinuous basins separated by low ranges with maximum summit levels near 1,500-1,800 m in eastern Nevada and Idaho. Mixed conifer forest lived at levels above 750-900 m, giving way to subalpine conifer forests near 1,500 m. These are the Eocene forests which contain taxa that are related to those that are now in the derived forests in California. However, the composition of the Eocene forests differed from the present ones in many ways.

Arcto-Tertiary Geoflora

History.—The Paleogene forests from British Columbia southward into Montana-Idaho include species of *Abies, Acer, Alnus, Betula, Chamaecyparis, Crataegus, Picea, Pinus, Populus, Pseudotsuga, Rhododendron, Sequoia, Thuja, Vaccinium* and others whose nearest descendants now contribute to the humid Coast forest. It dominates the coastal strip from central California north into Alaska, reaching inland to the northern Cascades and the panhandles of Idaho-British Columbia. The Paleogene forests of the northern cordilleran region also had numerous taxa whose nearest allies are no longer in the western United States. Some occur now in the eastern United States (an East American Element), notably species of *Acer, Aesculus, Betula, Carpinus, Diospyros, Fagus, Ilex, Liquidambar, Sassafras,* and *Ulmus.* Others are now only in eastern Asia (an East Asian Element), notably *Cercidiphyllum, Ginkgo, Keteleeria, Metasequoia, Pseudolarix,* and *Ulmus.* In addition, there were widely distributed species of *Abies, Acer, Betula, Crataegus, Picea, Pinus, Populus* and other dicots with their nearest relatives in eastern Asia. Emphasis must be placed on the fact that species of all these taxa were regularly admixed in rich communities, far richer than any which have survived.

Paleogene forests show an important change in composition south of the latitude of central Idaho (~ 44° N). The conifers and their associated hardwoods are not preponderantly those of very humid requirements, and together with their associates which usually do not occur farther north, suggest warmer and sunnier conditions. These forests, composed of *Abies, Picea, Pinus,* and *Pseudotsuga* show more affinity with the modern Sierran and Rocky Mountain forests than those of the Coast, for they include species of *Acer, Cercocarpus, Philadelphus, Ptelea, Ribes,* and *Salix* allied to those now in the central Rocky Mountain forest. Like the mesic forests farther north, they also have taxa allied to those now in the eastern United States and eastern Asia. By contrast, they have a number of sclerophylls, in *Berberis* (*Mahonia*), *Cercocarpus, Quercus, Rhus* subg. *Schmaltzia, Vauquelinia,* and *Zizyphus* that now reach up to the forest margin in New Mexico, Arizona, and southward. There also is a group that is now chiefly in the mountains of central Mexico (a Mexican Element), for instance *Astronium, Bursera, Cardiospermum, Colubrina, Oreopanax, Persea, Quercus,* and *Thouinia.* Since all these taxa that are no longer native to the central Rocky Mountain area now live under mild winter temperature and heavy summer rainfall, it is inferred that they were eliminated from the region as climate was modified following the Paleogene.

The mixed conifer and subalpine forests shifted coastward during the middle Tertiary as the cooling and drying trend accelerated. They displaced the more mesic mixed deciduous forests that had retreated earlier from the interior toward the coast, an area where low evaporation and mild temperature compensated for decreasing summer rain. The Miocene forests show a distribution pattern similar to the older floras, with those from central Oregon-Idaho and northward including more mesic taxa than those to the south. This is exemplified by the distribution of *Sequoia* and *Sequoiadendron* and their associates in the Neogene (fig. 2). The Miocene forests were exceptionally rich in taxa. This was not due solely to the presence in them of genera that are now in distant areas with summer rain (East Asian and East American Elements). These forests included many more conifers than occur in the descendant living forests. They represent two major groups. First, there were fossil species of *Abies, Chamaecyparis, Larix, Picea,* as well as dicots such as *Acer, Betula, Cornus, Populus* and others, that are now in eastern Asia and the eastern United States. Second, these forests were composed of species whose nearest descendants have since been segregated into the Coast, Sierran, and Rocky Mountain forests in the western United States. For example, in Oregon and Idaho *Abies* [*grandis* (Dougl.) Lindl.] and *Sequoia* are recorded together with *Picea* [*glauca* (Moench) Voss] , *P.* (*engelmannii* Parry), *Pinus* (*ponderosa* Laws.), and *Tsuga* [*mertensiana* (Bong.) Carr.] . In Nevada and southern Idaho, *Sequoiadendron* regularly occurs with species of *Picea* (*breweriana* S. Wats.), *Chamaecyparis* [*lawsoniana* (A. Murr.) Parl.] , and *Tsuga* (*mertensiana*). In Nevada, the remains of *Abies* (*bracteata* D. Don) are associated with *Abies* [*concolor* (Gord. & Glend.) Lindl.] , *A.* (*magnifica* A. Murr.), *Chamaecyparis* (*lawsoniana*), *Picea* (*breweriana*), *Sequoiadendron,* and *Populus* (*tremuloides* Michx.), taxa which are far removed from it today (fig. 3; Axelrod, 1976a). Also present are species that are scarcely separable from taxa that live with *Abies bracteata* today, notably *Arbutus menziesii* Pursh, *Lithocarpus densiflorus* (H. & A.) Rehd., *Quercus chrysolepis* Liebm., and *Q. wislizenii* A. DC. They contribute not only to the understory of the mixed conifer forest, but to a typical widespread broadleaved sclerophyll forest as well.

Clearly, the diversity of these forests was due to the mingling of taxa whose derivatives have since been restricted to narrower adaptive zones (see Axelrod, 1976b, fig. 4; 1977, figs. 5, 9). The species that occur now chiefly in the subalpine forest of the Sierra Nevada were represented then by ecotypes that contributed to mixed conifer forest well below the subalpine zone. Also, taxa now in the mixed evergreen forest at lower levels ranged well up into the mixed conifer forest, especially on warmer, south-facing slopes. These mixtures resulted from adequate rain in summer which reduced drought-stress at lower levels for subalpine taxa, and from mild winter temperatures that enabled some sclerophylls to range well up into the mixed conifer forest to near the subalpine zone (see Axelrod, 1976b, fig. 7).

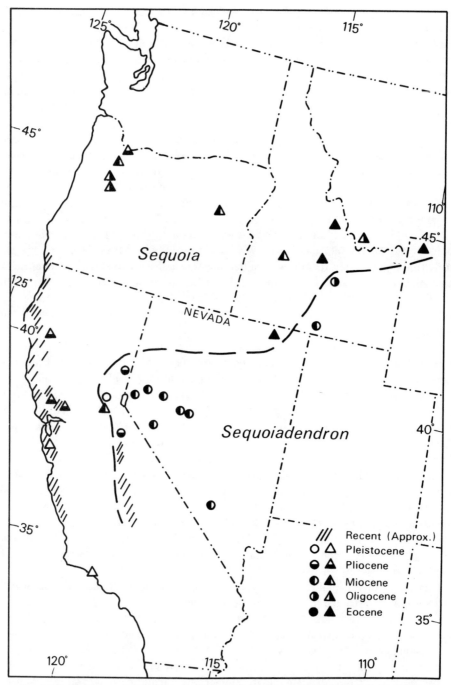

Fig. 2. The areas occupied by *Sequoia* and *Sequoiadendron* during the Tertiary were mutually exclusive. *Sequoia* was found farther north or coastward in a region of moister, milder climate, and was much more widely distributed. All known fossil localities for *Sequoiadendron* are shown on this map. For detailed modern distribution maps, see Griffin and Critchfield (1972).

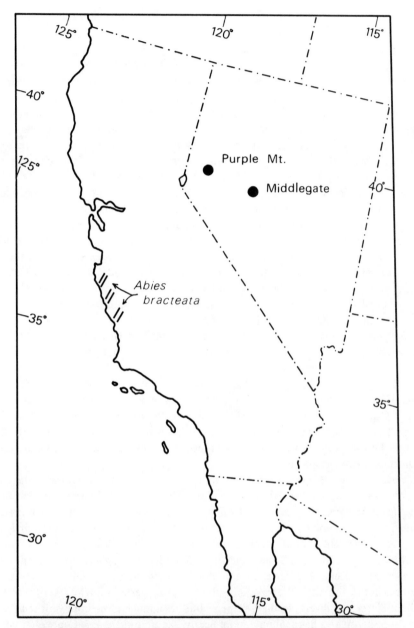

Fig. 3. Fossils similar to *Abies bracteata* are now known from the Miocene of western Nevada. Previously reported fossil records of the taxon represent other alliances (Axelrod, 1976a). A detailed map of modern distribution of *A. bracteata* is in Griffin and Critchfield (1972).

The mixed conifer and subalpine forests shifted farther coastward following the Middle Miocene as the cooling and drying trend continued. Open grassland and steppe replaced the retreating forests over the lowlands in the north, whereas juniper-pinyon-oak woodland rose to dominance over the Great Basin province in the Early Pliocene (10 m.y.). As the forests were restricted progressively coastward and into the mountains, they became more impoverished because taxa of the East American and East Asian Elements were being eliminated as summer rainfall decreased. Furthermore, decreasing summer rainfall now appears to have confined many conifers and associated species to higher, cooler levels where drought-stress was reduced. In addition, increasing cold was confining sclerophyllous taxa to the lower warmer levels in the surviving mixed conifer forest, where they were also largely removed from physical damage resulting from heavy snowfall (Axelrod, 1976a).

Regional differences in forests which were already apparent in the Oligocene, and were accentuated during the Neogene, became more marked during the Quaternary. The major differences in the present forests of California finally emerged during the Xerothermic, extending from about 8,000 to 4,000 years ago. Its effects were manifold (Axelrod, 1966, p. 45-55). In the Sierra Nevada, the spreading warm, dry climate finally shaped the present distribution of *Sequoiadendron*, which earlier had been disrupted by glaciation, restricting it to the interfluves between the major canyons. In addition, a number of taxa, for instance *Arbutus, Corylus, Lithocarpus, Pseudotsuga,* and *Taxus,* extend southward only into the central part of the range. Increased summer temperature and lowered precipitation southward presumably are inimical to seedling establishment in areas farther south. In the same way, subalpine forest loses a number of dominants in southern Tulare County, notably *Abies magnifica, Pinus albicaulis* Engelm., *P. monticola* Dougl., and *Tsuga mertensiana.* A similar pattern is shown by species of the Coast Forest, for *Abies grandis, Picea sitchensis* (Bong.) Carr., and *Tsuga heterophylla* (Raf.) Sarg. have only relict, isolated occurrences south of Cape Mendocino, chiefly in sheltered canyons close to the coast southward to Sonoma County. The area farther south presumably is now too dry in summer for them, for Coast Forest species covered the coastal region southward to the San Francisco Bay area into the early Pleistocene.

In view of the vicissitudes of the Quaternary, the question arises as to why the Klamath-Siskiyou region supports the richest forest flora. The area not only provides a haven for relicts (*Picea breweriana, Quercus sadleriana* R.Br. Campst., *Kalmiopsis*), but it is rich in conifers, with a total of 16-18 species contributing to the mixed conifer and subalpine forests in local areas. The conifers have been added to the region during the later Cenozoic, with the latest additions—e.g., *Picea engelmannii, Abies lasiocarpa* (Hook.) Nutt.—probably during the Pleistocene. The persistence of all of them in this region reflects the climate, not so much the diversity of terrain and substrate as suggested by Whittaker (1960): the northern Sierra is as diverse in these respects but has a much more impoverished flora. The Klamath-Siskiyou area has a climate more nearly like that of the late Cenozoic than any other part of the region. The precipitation season is longer than elsewhere in California, there is more summer rain, winters are not excessively cold, temperatures are more moderate in summer, and drought stress is lower here than in the Sierra and southward where the forests are not so diverse. In view of these conditions it is not surprising to find that species that are "typical" of the subalpine zone of the Sierra, as *Abies magnifica, Pinus contorta* Dougl. ex Loud. var. *latifolia* Engelm. (*P. murrayana* Grev. & Balf.), *P. monticola* Dougl., *Populus tremuloides,* as well as *Picea breweriana* and many of their associates (e.g.,

Juniperus communis L., *Pinus jeffreyi* Grev. & Balf., *Quercus vacciniifolia* Kell.), also contribute to mixed conifer forest in the Klamath-Siskiyou region. While it is true that the latter species often occur on serpentine at lower elevations, these are relict sites where they survive at low altitudes which they reached in the last glacial (Wisconsin), and are now removed from competition with the nearby flora (see p. 77). The association of these "subalpine" species with mixed conifer forest taxa parallels the relation shown by their fossil equivalents in the Neogene floras. Furthermore, the mixed conifer forest in the Klamath-Siskiyou region has evergreen dicots well into its upper part, also simulating the composition of Miocene and Pliocene floras. Lacking, of course, are the numerous taxa that are exotic to the Far West, and which were gradually eliminated here as summer rains decreased.

Nonetheless, there are still traces of these ancient links in the region. *Quercus sadleriana* is a member of the *Prinoideae*, now in eastern Asia and the eastern United States. *Pinus albicaulis* Engelm. is the only American member of the subsect. *Cembrae* whose other species occur in eastern Asia and the Alps-Carpathians. *Berberis (Mahonia) nervosa* Pursh is the sole member of sect. *Orientales* in North America, a group represented by scores of species in Asia. *Picea breweriana* belongs to sect. *Omorika* which is otherwise Eurasian. Other prominent taxa which are not restricted to the Siskiyou-Klamath region, but also have Asian affinities, include *Lithocarpus*, represented by about 300 species in Asia; *Chrysolepis*, a paired genus with *Castanopsis*, an Asian entity with about 120 species; and *Acer circinatum* Pursh, the sole American representative of the otherwise Asian sect. *Integrifolia*. To these we can add *Taxus* and *Torreya*, both discontinuous to Florida and occurring also in eastern Asia. *Calycanthus* and *Dirca* of central California occur elsewhere only in the eastern United States. Let us not forget that Asa Gray (1859) was aware of the numerous links between the California flora and those of eastern Asia and the eastern United States more than a century ago, not only the trees and shrubs, but numerous forest herbs as well.

In contrast to the forests elsewhere, the central Coast Ranges have only discontinuous stands, often of a single species. The fossil record shows that these forests were more continuous, and also richer in composition, during the moister phases of the Quaternary. They were confined to progressively higher elevations during the warm dry interglacials, and especially during the recent Xerothermic at which time the more mesic species disappeared from these comparatively low ranges. The severe drought at this time in southern California, as shown by the coastward spread of taxa from the interior (Axelrod, 1966b), probably accounts for the rather impoverished forests there, and also for the discontinuous distribution of numerous forest taxa, as listed by Munz (1935).

The pure forests of *Pinus ponderosa* Dougl. ex P.C. Laws., *P. jeffreyi*, and *Pseudotsuga menziesii* (Mirb.) Franco have developed at the borders of the more diverse, mixed conifer forests quite recently. Emphasis must be placed on the idea that they now occupy wholly new environments. It was the opening up of these new subzones during the Quaternary that provided a wealth of opportunities for radiation. This enabled numerous plant genera to proliferate new species adapted to these narrower, and certainly more unstable, subzones where extinction rather than persistence seems more likely to occur as moister climates appear again. Stebbins and Major (1965) have particularly emphasized the critical role of such ecotonal situations in the origin of the recently derived endemics of California.

Analysis.—Many of the genera of plants of California have wide distributions in the temperate regions of Eurasia and North America or are clearly derived from taxa that have had such a derivation (table 4). Other taxa are restricted to North America at present but are associated with the mesic derivatives of the Arcto-Tertiary Geoflora; these are also listed in table 4. To the seed plants listed in table 4 should be added the species of *Equisetum* and *Lycopodium* found in California and *Selaginella* as well. All ferns except some Pteridaceae, which are chiefly associated with more xeric vegetation types, Marsileaceae, and *Azolla* likewise appear to belong here. As we have seen, California is one of the most important areas of survival and persistence of relics derived from the great northern temperate Arcto-Tertiary forests (Wood, 1972; Wolfe, 1975) because of its equable climate and, since Late Pliocene time, the isolation of coastal areas from spreading aridity from the interior by the uplift of the Sierra-Cascade axis, and the Transverse and Peninsula ranges of southern California and Baja California.

Many genera and groups of genera basically associated with the forests of the north temperate regions have radiated extensively in California, producing clusters of species. This radiation has generally occurred in the drier, bordering vegetation of Madrean derivation. The dynamic relationship between the two sorts of vegetation has provided numerous opportunities for evolution, and has been stressed by many authors. Among the groups listed in table 4, in which speciation in California is evident, are Polemoniaceae-Polemonieae (see p. 52); the Scrophulariaceae-Rhinantheae (*Castilleja, Cordylanthus, Ophiocephalus, Orthocarpus, Pedicularis*); Asteraceae-Lactuceae (*Agoseris, Microseris, Nothocalais, Phalacroseris*); *Agrostis* (Crampton, 1961); *Allium*; *Aster*; *Calamagrostis* (Crampton, 1961); *Calochortus*; *Calystegia* (Brummitt, 1974); *Campanula*; *Carex*, especially sect. *Ovales*; *Cirsium* (Ownbey, Raven, and Kyhos, 1976); *Delphinium* (Lewis and Epling, 1954); *Dicentra* (Stern, 1961); *Dodecatheon*; *Erigeron*; *Eryngium*; *Gnaphalium*; *Iris*; *Juncus*; *Lathyrus*; *Lithophragma*; *Lomatium* (Theobald, 1966); *Lonicera*; *Perideridia* (Chuang and Constance, 1969); *Pleuropogon* (Crampton, 1961); *Poa* (Crampton, 1961); *Polygonum* sect. *Duravia* (Hedberg, 1946; Mertens and Raven, 1965); *Potentilla* and its derivatives *Horkelia, Ivesia,* and *Purpusia*; *Ribes*; *Sanicula* (Shan and Constance, 1951; Bell, 1954); *Senecio*; *Silene* (Hitchcock and Maguire, 1947); *Stachys*; *Stipa* (Crampton, 1961); *Tauschia*; and *Viola*. There are approximately 398 genera and 2454 species included among the taxa listed in table 4. They comprise about 46% of the genera and 48.6% of the species in the State. In the California Floristic Province, more than half of the genera and species have Arcto-Tertiary affinities.

Only about 280 (12%) of these species are annuals, in contrast to the 28.6% in the flora of the State as a whole. In relatively few genera of clear northern affinities does the evolution of annual species appear to have taken place in or near California to any extent—Asteraceae-Lactuceae, Boraginaceae, Polemoniaceae-Polemonieae, and Scrophulariaceae-Rhinantheae, *Agrostis, Bromus, Campanula, Hesperocnide, Lupinus,* and *Polygonum* sect. *Duravia* (Hedberg, 1946; Mertens and Raven, 1965), are examples. The monocots in this group number about 79 genera and 646 species, the latter comprising about 30% of the total, considerably higher than for the flora as a whole but more or less consistent with the proportion of monocots in other north temperate floras.

Madro-Tertiary Geoflora

History.—Fully a third of the non-desert part of California is (or was) covered with sclerophyll vegetation and related, derivative communities. Fossil plants representing live oak woodland and associated sclerophyll vegetation first appear in the Middle and Late Eocene,

TABLE 4

North Temperate Seed Plants in the California Flora[1]

Aceraceae

Apiaceae (including many widespread genera and significant endemic groups in *Eryngium, Lomatium, Tauschia, Sanicula,* and *Perideridia*; Chuang and Constance, 1969; Giannasi and Chuang, 1976)

Asteraceae-Lactuceae, subtribe Microseridinae (*Agoseris, Apargidium, Microseris, Nothocalais, Phalacroseris*; Feuer and Tomb, 1977)

Betulaceae

Boraginaceae, see pp. 50-51.

Brassicaceae [*Arabis, Barbarea, Cardamine,* (including *Dentaria*), *Descurainia, Draba, Erysimum, Hutchinsia, Nasturtium, Rorippa, Smelowskia, Subularia*]

Caprifoliaceae

Caryophyllaceae (10 genera, 61 species in CFP[2], of which 38 endemic; large centers of differentiation in *Arenaria* sens. lat. and in *Silene,* with 500 species, of which 54 occur in North America according to Hitchcock and Maguire, 1947; the largest center of differentiation is in the Mediterranean region and parallels the much smaller one in and around Calif. Annual species have evolved in both regions.)

Crassulaceae (*Crassula, Sedum*)

Cupressaceae (except *Cupressus*)

Cyperaceae (*Carex, Dulichium, Eriophorum, Rhynchospora,* together with some species of *Eleocharis* and *Scirpus. Carex* is the largest genus in Calif., with 142 native species, but including only 22 endemics.)

Droseraceae

Empetraceae

Ericaceae except Arbuteae and perhaps *Gaultheria*

Fabaceae-Faboideae (*Glycyrrhiza; Lathyrus,* Hitchcock 1952; *Oxytropis; Thermopsis;* and *Vicia,* with the parallel evolution of annual species in the Old and New World)

Fagaceae (except perhaps for some groups of *Quercus* mentioned in table 5)

Gentianaceae (*Frasera, Gentiana* sens. lat., *Swertia*)

Juglandaceae

Juncaceae (47 species of *Juncus,* 8 of *Luzula,* with respectively 11 and 1 endemic to CFP)

Liliaceae (Includes many North American endemic genera and species; a complex family in which the relationships are poorly understood, and in which it is therefore very difficult to analyze patterns of distribution; cf. Raven and Axelrod, 1974. *Allium, Calochortus, Chlorogalum,* and *Odontostomum* will be discussed separately.)

Menthaceae (*Agastache, Lycopus, Mentha, Prunella, Pycnanthemum, Satureja, Scutellaria, Stachys*)

Nymphaeaceae

Orchidaceae (10 genera and 22 species, mainly widespread and only 1 endemic to CFP)

Orobanchaceae (Heckard, 1973)

Pinaceae (with early radiation into Madrean vegetation by *Pinus* subsects. *Cembroides* and *Oocarpae*; Little and Critchfield, 1969)

Poaceae (*Agropyron, Agrostis, Alopecurus, Beckmannia, Bromus, Calamagrostis, Cinna, Danthonia, Deschampsia, Elymus, Festuca, Glyceria, Hesperochloa, Hierochloë, Hordeum, Hystrix, Koeleria, Leersia, Melica, Oryzopsis, Phalaris, Phleum, Phragmites, Pleuropogon, Poa, Puccinellia, Scribneria, Sitanion, Spartina, Sphenopholis, Stipa, Trisetum, Vulpia*)

Polemonaceae, tribe Polemonieae (see p. 52)

Polygonaceae (*Oxyria, Polygonum, Rumex,* with a center of differentiation in drier habitats in one group of *Polygonum*)

Primulaceae

Pyrolaceae

Ranunculaceae (13 genera and 81 native species in Calif., 27 in CFP, and 36 of these endemic. All genera widespread in temperate regions of the Northern Hemisphere; one distinctive group of *Delphinium* in dry habitats in and near California—Lewis and Epling, 1954.)

---

[1] In this table, "endemic" refers to the California Floristic Province unless otherwise specified.

[2] California Floristic Province.

TABLE 4 (continued)
North Temperate Seed Plants in the California Flora

Rosaceae (*Agrimonia, Amelanchier, Aphanes, Aruncus, Chamaebatiaria, Crataegus, Fragaria, Geum, Holodiscus, Luetkea, Malus, Petrophytum; Potentilla,* as well as its western North American derivatives *Horkelia,* 17 species, *Ivesia,* 22 species, and *Purpusia,* 1 species; *Prunus, Rosa, Rubus, Sanguisorba, Sibbaldia, Sorbus, Spiraea,* as well as the more local monotypic genera *Heteromeles* and *Oemleria*)

Salicaceae

Saxifragaceae (except Hydrangioideae, the remainder comprising 16 genera and 92 species in the CFP, of which 40 are endemic and *Heuchera, Ribes,* and a few other genera fairly diverse)

Scrophulariaceae-Rhinantheae (*Castilleja,* 200 species, primarily North American, almost all western, with secondary radiation in South America, 2 species in Siberia, 33 in Calif.; *Cordylanthus,* 18 species, 16 in Calif., with 10 in CFP, 9 endemic, the others in deserts, Chuang and Heckard, 1973; Heckard, personal communication; the monotypic *Ophiocephalus,* endemic to Sierra San Pedro Mártir of Baja Calif., Chuang and Heckard, 1973; *Orthocarpus,* 25 species, mostly in Calif., 1 in Andes; and *Pedicularis,* 500 species, 10 in Calif., 4 endemic)

Taxaceae

Urticaceae

Achillea (Asteraceae-Anthemideae: 200, mostly Eurasian, 1 in Calif.; Tyrl, 1975)

Achlys (Berberidaceae: 2, 1 each in E. Asia and Pacific North America)

Adenocaulon (Asteraceae-?Mutisieae: 4, North Temperate to Chile, 1 in Calif.)

Allium (Amaryllidaceae: 450, Northern Hemisphere, mostly in the Old World but well represented in North America and especially in the W. United States; 38 species in Calif., 31 in CFP, of which 24 are endemic. Thus there is a major site of differentiation within the Madrean vegetation of Calif., similar to that of the Mediterranean region itself and the Near East.)

Anaphalis (Asteraceae-Inuleae: 35 North Temperate, 1 in Calif.)

Antennaria (Asteraceae, Inuleae, 100, temperate zones, 10 in Calif., 1 endemic in Siskiyou Mts.)

Aralia (Araliaceae: 35, Asia and North America, 1 in Calif., endemic)

Arceuthobium (Viscaceae: 28, 9 in Calif., 2 endemic, considerable radiation in Mexico and the W. United States; Hawksworth and Wiens, 1972)

Armeria (Plumbaginaceae: 80, North Temperate and Andes, ours northern)

Arnica (Asteraceae-Senecioneae: 32, North Temperate, 18, including 4 endemics, in Calif., as well as the monotypic endemic *Whitneya*)

Artemisia (Asteraceae-Anthemideae: 400, mainly North Temperate, 15, including 5 endemics, in Calif.)

Asarum (Aristolochiaceae: 70, 3 in Calif., 2 endemic)

Aster (Asteraceae, Astereae: 250, widespread in temperate regions, with a number in Madrean vegetation)

Berberis (Berberidaceae: 500, Asia and W. North America to Central America, 13 species in Calif., 11 in CFP, of which 8 endemic)

Calocedrus (Cupressaceae: 3, 1 in Calif., 2 in Southeast Asia and Taiwan)

Calochortus (Liliaceae, but very distinct: 60, temperate W. North America to Guatemala, 38 in Calif. and with a major center of differentiation in Madrean vegetation)

Calycanthus (Calycanthaceae: 3, 1 in Calif., 2 in E. United States)

Calystegia (Convolvulaceae: 52 species and subspecies, more than half in CFP, an important center of evolution; Brummitt, 1974)

Campanula (Campanulaceae: 250, largely North Temperate and of the Old World, 9 in Calif. including 6 endemics, 4 perennials and 2 annuals; Heckard, 1969)

Celtis (Ulmaceae: 80, Northern Hemisphere and South Africa, 1 reaching Calif.)

Cercis (Fabaceae-Caesalpinioideae: 7, 2 in North America)

Chrysanthemum [Asteraceae-Anthemideae: 200, Eurasia, N. Africa, a few in North America and 1 in coastal Calif., *Chrysanthemum huronense* (Nutt.) Hultén, including *Tanacetum camphoratum* Less. and *T. douglasii* DC.]

Circaea (Onagraceae: 8, North Temperate, 1 in Calif.)

Cirsium (Asteraceae-Cynareae: about 300 North Temperate species, half in Eurasia, 28 in Calif., including 16 endemics—an important center of diversification of the genus, which is also well represented in the drier regions of W. United States and in Mexico; Ownbey et al., 1976)

TABLE 4 (continued)
North Temperate Seed Plants in the California Flora

Corydalis (Fumariaceae: 320 North Temperate, 1 in East Africa; 10 in North America, 2 in Calif. Own-bey, 1947)

Crepis (Asteraceae-Lactuceae: 200, mainly Eurasian, with 2 groups in North America, represented by 9 species in Calif.)

Cuscuta (Cuscutaceae: 170, cosmopolitan, 14 in Calif., 2 endemic)

Darlingtonia (Sarraceniaceae: 1, N. Calif. to N. Cent. Ore.; family restricted to Western Hemisphere)

Dicentra (Fumariaceae: 19, Asia and Temperate North America [Stern, 1961] with 6 in Calif., 4 en-demic to CFP, including the most primitive section, sect. *Chrysocapnos*)

Dirca (Thymelaeaceae: 2, 1 endemic to San Francisco Bay region, the other widespread in Cent. and E. United States; presumably ancient Arcto-Tertiary genus)

Epilobium (Onagraceae: 200, worldwide, 20 in Calif., 2 endemic)

Erigeron (Asteraceae, Astereae: 200, widespread in temperate regions, 44 in Calif., 35 in CFP, 11 en-demic)

Euonymus (Celastraceae: 175, North Temperate, 1 in Calif.)

Euphorbia subg. *Esula* (Euphorbiaceae: 500, worldwide but mainly North Temperate, 4 in Calif., 2 perennials, 2 annuals; Webster, 1967)

Fraxinus (Oleaceae: 70, 3 in CFP, of which 2 endemic)

Galium (Rubiaceae: 400, cosmopolitan. Of the approximately 30 native species in Calif., perhaps 6 might be of northern affinities; the ditypic *Kelloggia* belongs here also; Dempster, 1975)

Geranium (Geraniaceae: 400, including many North Temperate perennial species; 7 native perennials and 2 native annuals in Calif.)

Gnaphalium (Asteraceae, Inuleae: 200, cosmopolitan, 9 in Calif., 4 endemic to CFP)

Heterotheca (Asteraceae, Astereae: 25, including *Chrysopsis*, widespread in temperate North America, 6 in Calif., 2 endemic in S. CFP)

Hieracium (Asteraceae-Lactuceae: 700, mainly North Temperate; 7 in Calif., with 3 endemic)

Iliamna (Malvaceae: 7, temperate North America, 1 endemic at N. edge of CFP, 1 mostly outside it in N.E. Calif. and adjacent Ore.; Wiggins, 1936. Perhaps ultimately derived from the Mexican, Cen-tral American, and West Indiana *Phrymosia*, and probably secondary in Arcto-Tertiary vegetation closely related to *Malacothamnus*; Bates, 1963.)

Iris (Iridaceae: 300, mainly in Eurasia, with 13, including 8 endemics, in Calif.; Lenz, 1958)

Keckiella (Scrophulariaceae: 7, CFP, 6 endemic, 1 to Ariz. and cent. Baja Calif. Although Cheloneae are clearly of W. North American origin, *Keckiella* appears to be a specialized group parallel to *Diplacus* that has been derived largely in Madrean vegetation.)

Lactuca (Asteraceae-Lactuceae: 100, mainly North Temperate, 2 in Calif.)

Leucophysalis [Solanaceae: 9, 7 in Asia, 2 in North America, including *L. nana* (A. Gray) Averett]

Lilaea (Lilaeaceae, monotypic family related to Juncaginaceae; mountains of W. North America and South America)

Linaria (Scrophulariaceae: 150, Eurasia and N. Africa, especially Mediterranean region, with 2 wide-spread and closely related species in North America, both in Calif.)

Linum s. str. (Linaceae: 2 species in Calif.)

Luina (including *Cacaliopsis*; Asteraceae-Senecioneae: 6, 2 in N.W. Calif.)

Lupinus (Fabaceae-Faboideae: 200, W. North America through the Andes, with about 3 species in E. United States and 6 annual species in Mediterranean region; whether they are derived from annual species in North America or independently evolved from perennial ancestors is uncertain. In Calif., 86 species, with 68 in CFP, 47 of them endemic. Genus occurs from moist forest habitats out onto the deserts. Dunn, 1971, has postulated that the species in the E. United States might have been derived from South American ancestors following long-distance dispersal from Brazil.)

Lysichiton (Araceae: 2, 1 in E. Asia, 1 in North America)

Matricaria (Asteraceae, Anthemideae: 50, 2 in Calif. and temperate North America generally, 40 in W. Eurasia and N. Africa, 10 in South Africa)

Nothochelone (Scrophulariaceae: 1, forests of W. North America S. to Siskiyou Mts., Calif.; Straw, 1966)

Oxalis (Oxalidaceae: 800, cosmopolitan, with 5 native species in Calif.)

Paeonia (Paeoniaceae: 33, Eurasia, except for 2 in Calif., 1 endemic)

TABLE 4 (continued)
North Temperate Seed Plants in the California Flora

Paxistima (Celastraceae: 1, forests of W. United States)

Penstemon (Scrophulariaceae: 250, N. America to Central America, especially W. United States; 50 in Calif., 32 in the CFP, of which 20 are endemic. This genus has evolved mainly in semiarid habits in the W. United States, S. into Mexico, and a number of species have doubtless entered Calif. since mid-Pliocene time; Straw, 1966.)

Petasites (Asteraceae-Senecioneae: 5, Northern Hemisphere, with 1 in Calif.)

Phyllospadix (Zosteraceae: 5 species of marine angiosperms of the N. Pacific, with 2 in Calif.)

Rhamnus (Rhamnaceae: 160, cosmopolitan, with 9 species in Calif., 3 in the probably valid genus *Frangula*, and 7 endemic)

Saussurea (Asteraceae-Cynareae: 400, all but 3 in Asia; the 1 North American species occurs in Calif.)

Scheuchzeria (Scheuchzeriaceae, a monotypic family of circumboreal distribution)

Senecio (Asteraceae-Senecioneae: 2000, with 33, including 15 endemics, in Calif., the genera *Cacaliopsis* and *Luina* derived within this complex)

Sequoia (Taxodiaceae: 1, endemic)

Sequoiadendron (Taxodiaceae: 1, endemic)

Shepherdia (Elaeagnaceae: 3, North America, 1 in Calif.)

Sidalcea (Malvaceae: 22, W. North America, 18 in Calif., including one derived group of 5 annual species endemic to Calif., related to European genera of the *Malva*-complex according to D. M. Bates, personal communication, and therefore almost certainly of Arcto-Tertiary origin, unlike most North American Malvaceae)

Solidago (Asteraceae, Astereae: 100, including 1 in Eurasia, 7 in CFP, 4 endemic)

Sparganium (Sparganiaceae: 20, North Temperate and Australia, with 3 or 4 in Calif.)

Sphaeromeria (Asteraceae-Anthemideae: 8, 2 in Calif., 2 in CFP, including 1 endemic to Sierra San Pedro Mártir, Baja Calif.; Holmgren, Schultz, and Lowrey, 1976)

Staphylea (Staphyleaceae: 10, North Temperate, 1 in Calif., endemic)

Styrax (Styracaceae: 130, mainly Eurasia, with about a dozen species in North America and 1 in California, supposedly conspecific with a Mediterranean one)

Synthyris (Scrophulariaceae: 14, 2 in Calif., 1 endemic)

Taraxacum (Asteraceae-Lactuceae: many species, mainly North Temperate, but poorly represented by native species in North America; 1 endemic in San Bernardino Mts. of Calif.)

Toxicodendron (Anacardiaceae: 15, E. Asia and temperate North America to Colombia, 1 in Calif.; Barkley, 1937; Gillis, 1971)

Valeriana (Valerianaceae: 200, mainly North Temperate, 3 in Calif.)

Vancouveria (Berberidaceae: 3, coastal Wash. to Calif.)

Veronica (Scrophulariaceae: 300, mainly North Temperate, 8 in Calif., 1 endemic)

Viola (Violaceae: 500, cosmopolitan, 21 in Calif. of which 8 endemic to CFP)

Vitis (Vitaceae: 60-70, Northern Hemisphere, 2 in Calif., both endemic)

---

in the central and southern Rocky Mountains. Here we find live oak, laurel, madrone, and pinyon pine associated with species of many genera that still live with them today. These include *Cercocarpus, Chamaebatiaria, Celtis, Platanus, Populus, Rhus* subg. *Schmaltzia, Salix,* and *Vauquelinia,* with drier sites supporting *Bursera, Caesalpinia, Colubrina, Euphorbia* subg. *Agaloma* (*californica* Benth.), *Prosopis, Tephrosia,* and *Zizyphus.* Clearly, this was a diverse woodland whose associates included both drought-resistant sclerophylls as well as drought-deciduous shrubs of subtropic affinity.

The record as now known shows that sclerophyllous taxa contributing to vegetation representing the Madro-Tertiary Geoflora were already in ecotone with the drier margins of the Arcto-Tertiary Geoflora at this early date (~ 50 m.y.). Floras in the interior show that mixed conifer forest reached down cooler, moister canyons from higher levels to interfinger with the Madrean vegetation of subhumid requirements that dominated over the warmer, drier lowlands. And today, derivative communities from these sources still form an ecotone over a wide area in the western United States. The record clearly implies that Madro-Tertiary taxa have had a long, still unrecorded pre-Eocene history. They probably originated earlier, on drier lee slopes and rocky outcrops over low-middle latitudes, a supposition consistent with the records of a few rare taxa in Paleocene floras of the southern Rocky Mountains that appear to represent them (Axelrod, 1958).

Madro-Tertiary vegetation spread with expanding dry climate following the Eocene. We do not yet know when it initially entered California because the record is very incomplete. Preliminary analysis of pollen in the Eocene Del Mar Formation by Dr. Wm. S. Ting, however, indicates sclerophylls were already present. The site, which is west of the San Andreas fault, was then situated about 400 km farther south, near the present site of El Rosario, Baja California. Sclerophyllous vegetation rapidly expanded with spreading dry climate and had already assumed dominance over interior southern California by the Miocene. The very rich Tehachapi and Mint Canyon floras clearly demonstrate that the present Mohave-Sonoran desert region formed a single floristic province at that time, one dominated by sclerophyllous vegetation and thorn forest (Axelrod, 1950).

The rich woodland was characterized by live oaks (cf. *Quercus agrifolia* Née, *Q. brandegeei* Goldm., *Q. emoryi* Torr., *Q. oblongifolia* Torr., *Q. wislizenii*), pinyon pine, *Cupressus*, laurels (*Persea, Umbellularia*), palms [*Brahea* (*Erythea*), *Sabal*], ironwood (*Lyonothamnus*), and madrone (*Arbutus*). It had a very rich understory of sclerophyllous shrubs, notably species of *Arctostaphylos, Berberis* (*Mahonia*), *Ceanothus, Cercocarpus, Heteromeles, Malosma, Prunus* subg. *Laurocerasus, Quercus, Rhamnus,* and *Rhus* subg. *Schmaltzia.* Many of the fossil species have their nearest relatives in southern California, but others are in the southwestern United States and Mexico where there is regular summer rain. These Miocene floras from southeastern California also have numerous taxa representing thorn scrub vegetation, distributed in *Acacia, Bursera, Caesalpinia, Cardiospermum, Colubrina, Dodonaea, Ficus, Randia, Zizyphus,* and others. All of them now attain optimum development south of the border, where climate is frostless or nearly so, and rain comes chiefly in the warm season. During the Miocene a low topographic barrier at the site of the present Peninsular Ranges separated this semiarid vegetation from the coastal strip. In that area, palm-oak-laurel forest and closed-cone pine forest dominated under a moister climate with lower evaporation rate. In the interior and to the north, there was a rapid rise in terrain near the present boundary between the Mohave and Great Basin desert provinces. In that higher, moister, and cooler region to the north, Madrean sclerophyll vegetation interfingered with the mixed conifer hardwood forests of the Arcto-Tertiary Geoflora (Axelrod, 1956, fig. 16; 1968, fig. 7).

During the Middle and Late Miocene this ecotone was chiefly one of live oak woodland marginal to a rich Sierran-type forest in which *Sequoiadendron* was abundant and widespread (fig. 2). Not represented in the Nevada province, however, are records of the subtropic thorn forest that dominated over the warmer and drier lowlands in southeastern California. As the trend to drier climate continued, the mixed conifer forest in the Great Basin region retreated to the west and into the mountains, and oak-pinyon-juniper woodland and associated sclerophyllous shrubs had assumed dominance over the lowlands by the Pliocene ($\sim$ 10 m.y.). Increasing cold and reduced summer rain in the Late Pliocene finally eliminated most of the sclerophylls over the region, leaving an impoverished pinyon-juniper woodland above the spreading dry steppe at lower levels.

Sclerophyll vegetation is recorded on the west Sierran slope in the Late Miocene, where it is in ecotone with deciduous hardwood forest which locally includes members of the Coast Forest, notably *Sequoia* and *Chamaecyparis.* Deciduous hardwoods that probably dominated in the moister valleys included species of *Aesculus, Carya, Liquidambar, Nyssa, Populus* that are no longer in the region, as well as sclerophylls that are now found in Mexico, as in *Berberis (Mahonia), Ilex, Karwinskia, Persea,* and *Ungnadia.* Live oaks are frequent and presumably covered the warmer and drier well-drained slopes, together with *Arctostaphylos, Ceanothus, Cercocarpus, Robinia,* and others. The exotics were largely eliminated from the low Sierran rise by the Middle Pliocene, which was the driest part of the Tertiary in this region.

At this time sclerophyll vegetation first assumed dominance in west-central California, displacing deciduous hardwood forests that had dominated there since the Miocene. Accompanying the rise to dominance of sclerophyll vegetation, we find the appearance of taxa which have since been restricted to insular and coastal southern California (figs. 4, 5), notably *Ceanothus (spinosus* Nutt.), *Lyonothamnus (floribundus* A. Gray), *Malosma* [*laurina* (Nutt.) Nutt. ex Abrams] , *Prunus (Laurocerasus) lyonii* (Eastw.) Sarg., *Rhus (Schmaltzia) ovata* S. Wats., as well as species of *Populus (arizonica* Sarg.), *Quercus* [*pungens* Liebm. var. *vaseyana* (Buckl.) C.H. Muller] , *Sapindus,* and others that have their nearest descendants in the region from Arizona to western Texas. Associated on nearby cooler, moister slopes was a dominant *Lithocarpus-Arbutus* sclerophyll forest which was in the Coast Ranges at an earlier date, as shown by its occurrence near Coalinga in the Miocene.

A few records indicate that the outer coastal strip in central California was covered with closed-cone pine forest in the Pliocene, as shown by fossil cones similar to those of *Pinus radiata* D. Don (fig. 6) and *P. muricata* D. Don. Pollen records indicate that this forest was already well developed in coastal southern California in the Miocene.

Exotic woodland taxa had largely been eliminated from the region by the Late Pliocene, except for the narrow coastal strip where high equability, low evaporation, and probably some summer rain still provided conditions suitable for them. As precipitation increased and temperature was lowered many sclerophyllous taxa were confined to southern California, chiefly by the process of extinction in their northern areas. Forests now spread more widely and displaced woodland as well as broadleaved sclerophyll forest over the lowlands. During the Quaternary, forests descended to lower levels at times of cold-moist (glacial) climate, and they also shifted well south of their present area. During the Early (or Middle?) Pleistocene in interior southern California mixed conifer forest descended fully 900 m below its present level, at which time it included taxa that are now chiefly coastal (*Arbutus, Populus trichocarpa* Torr. & Gray—oval-leaf type) as well as relicts that indicate summer rain (*Acer brachypterum* Woot. & Standl., *Magnolia*). In the Late Pleistocene, redwood forest shifted to Carpinteria, 240 km south of its present southern limit. Data are still insufficient to estimate the Pleistocene lowering of forest in the Sierra, though relict stands on favorable north slope sites at low altitudes indicate considerable lowering did take place. These relict stands provide only a general measure of the lowering of the forest belt because forest patches that were even lower must have been eliminated by the warmer, drier climate during the Xerothermic period; others were no doubt removed by early settlers.

Cold and wet periods that accompanied the glacial ages were succeeded by warm-dry ones. It was the last one, the Xerothermic period, that shaped many of the present distri-

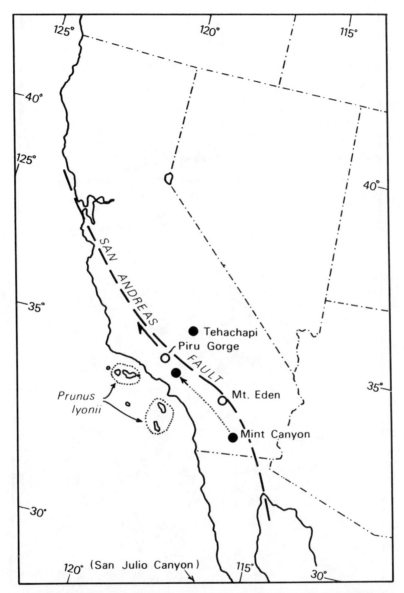

Fig. 4. *Prunus* (*Laurocerasus*) *lyonii* now occurs on some of the islands off southern California and also locally in the deep arroyos of south-central Baja California. A similar evergreen cherry occurs in Miocene (●) and Pliocene (○) floras on the mainland. Note that the Mint Canyon flora has been displaced northwest about 300 km. The Pliocene floras have been displaced about one-half as much, but their earlier positions are not shown here for reasons of map clarity.

butions (Axelrod, 1966b). The spread of warmer, drier climate enabled taxa of semidesert sites to reach the coast in southern California, the southern or Catalinan islands, and also the hot, dry valleys of the inner south Coast Ranges. In addition, chaparral species typical of southern California dispersed northward into the central Coast Ranges and the southern

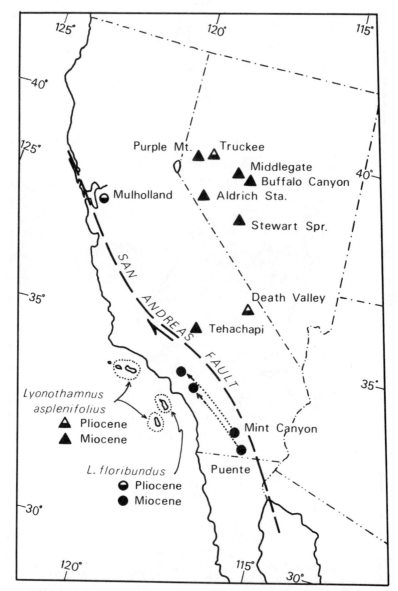

Fig. 5. Fossils allied to the insular *Lyonothamnus asplenifolius* and *L. floribundus* occur on the mainland. Taxa similar to *L. floribundus* are chiefly of near-coastal
occurrence, whereas those allied to *L. asplenifolius* are largely of interior distribution.
The Miocene Mint Canyon and Puente localities have been displaced northwest about
240 km by movement of the block west of the San Andreas fault.

Sierra. It probably was at this time that *Pinus sabiniana* Dougl. woodland was eliminated
from southern California, a region where it was well established in the Pliocene. Evidence
of the recent spread of dry climate is seen in the relict stands of this woodland well up in
the Sierra Nevada where they are surrounded by mixed conifer forest, and also in areas
well toward the coast in the North Coast Ranges. Spreading dry climate during the Xero-

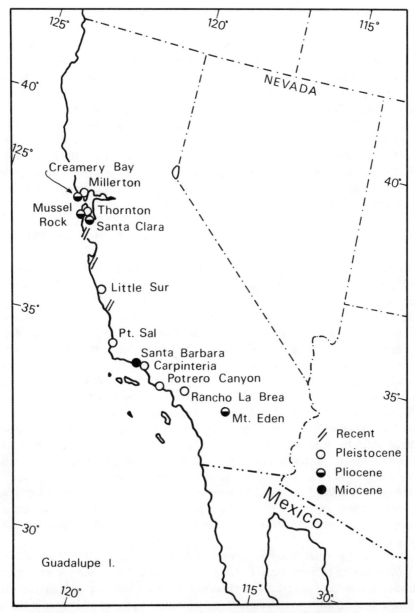

Fig. 6. Showing the present (///) and Pleistocene distribution of *Pinus radiata*, and that of its close relative, *P. lawsoniana*, in Pliocene and older rocks.

thermic also appears to account in part for disruption of the *Sequoiadendron* forest. In addition, the continuous closed-cone pine forest that covered the coastal strip (fig. 6) was fragmented. The present relict stands occur chiefly on or near coastal highlands where there is heavy fog in summer and a strong upwelling of colder subsurface coastal water. The evidence as now known implies that all vegetation units in California, from the coastal strip up to timberline (Taylor, 1976), were affected by the spread of drier, warmer climate during the Xerothermic.

The survival of a number of Tertiary woodland relicts on the coastal strip of southern California and Baja California, as well as on the islands (figs. 4-6), parallels in some ways the history of forest relicts in the Siskiyou Mountain region. It is the insular area and near-by coastal strip that has the most equable climate (M 65+), and where drought stress is re-duced in summer. This is where relict taxa occur that have their nearest allies in the wood-lands of central Mexico, a region also of high equability (M 60+), of summer rainfall, and little or no frost—a climate like that inferred for the Miocene and Pliocene of California. Among the links between insular and coastal southern California and the mountains of Mex-ico are *Arbutus menziesii-A. xalapensis* HBK., *Ceanothus arboreus* Greene-*C. coeruleus* Lag., *Cercocarpus traskiae* Eastw.-*C. mojadensis* C. Schneid., *Comarostaphylis, Garrya, He-teromeles, Myrica californica* Cham. & Schlecht.-*M. mexicana* Willd., *Pinus muricata-P. radi-ata-P. remorata* H.L. Mason and *P. oocarpa* Schiede and its allies (Critchfield, 1967), *Prunus lyonii-P. prionophylla* Standl., and *Vaccinium ovatum* Pursh-*V. confertum* HBK. As the land area west of the San Andreas fault moved north, it brought some of these Madrean taxa from southern California into juxtaposition with the Coast forest of Arcto-Tertiary affinity. This evidently was in the Late Pliocene as judged from the record as now known in central California.

Like the forests, the record shows that typical members of two or more present-day woodland communities were associated well into the Pliocene (Axelrod, 1977, fig. 10). Thus, the present woodlands of California were segregated into associations that are con-fined now to more localized environments which have developed here since the Pliocene. The insular region has a unique woodland filled with relicts. The outer coast is character-ized by an evergreen live oak woodland which gives way inland to *Pinus sabiniana* wood-land-savanna with deciduous oaks where frost frequence rises to more than one percent of the hours of the year. The inner valleys of southern California have a *Quercus engelmannii* woodland which has relict outposts to the west (Claremont, Glendora, Pasadena). These were disrupted by the intrusion of *Juglans*-live oak woodland into the area when more equable climate returned following the Xerothermic period (see Axelrod, 1977, figs. 8, 9). As these major woodland communities were segregated into more restricted topographic-climatic areas, the associated herbaceous and perennial flora also responded to the new en-vironments. The result is seen now in the diverse, related species in numerous genera, as outlined below.

Coastal sage vegetation, a community dominated by species of *Artemisia* (*californica* Less.), *Haplopappus* (*Ericameria, Hazardia*), *Lotus, Salvia, Yucca,* and others, assumed dominance as woodland retreated from lower, drier levels. The taxa were already present as regular members of the woodland-savanna vegetation, and they still occur in the lower, drier parts of the oak-savanna belt today. Coastal sage is a new climax community, the taxa of which were already adapted to the drier climate that developed below the savanna-wood-land zone. This vegetation, which has not yet been described adequately, includes several very distinct associations (Diegan, Venturan, Morroan, Franciscan, Coalingan), some of which (Diegan, Venturan) have already largely disappeared under spreading urbanization.

Much the same can be said for chaparral, which has reached its maximum extent under the influence of man's activities. The shrubs, derived chiefly from a rich woodland mosaic, were preadapted to the rapidly rising mountain slopes in the Late Pliocene-Early Pleistocene, to the dry summers that emerged following the pluvials, and to the raging fires that tend to perpetuate them. Several distinct associations are present in southern California and elsewhere, yet none of them is adequately described. One of them, that dominated by chamise (*Adenostoma fasciculatum* H. & A.), covers thousands of acres as a pure association. It appears to be of very recent origin at the lower margin of woodland, giving way to coastal sage in still drier sites. In southwestern Oregon, the present distribution of chaparral reflects its expansion during the Xerothermic period (Detling, 1961; 1968) and the migration of xerophytic communities into eastern Washington-northern Idaho at the same time has been discussed by Daubenmire (1975).

It is amply clear that the segregation of new woodland, chaparral and sage communities into new, narrower environments was accompanied by a decrease in the diversity of the woody communities. Segregation was determined chiefly by the degree of equability of climate in these diverse semiarid and subhumid regions. As these new, more local and more specialized environments appeared, they provided opportunities for explosive speciation of diverse taxa, chiefly among the annual dicots.

Analysis.—The taxa that are primarily associated with vegetation types derived from the Madro-Tertiary Geoflora in California comprise about 196 genera and 1460 species—roughly a quarter of the genera and a third of the species in the California Floristic Province (tables 5, 6). They are known or presumed to have had long histories in association with the sclerophyll vegetation of western North America. Approximately 55% of these species are annual, indicating the strong evolutionary radiation of many genera such as *Clarkia, Cryptantha, Hesperolinon, Lasthenia, Lupinus, Mimulus,* and *Phacelia* which have probably been associated since their origin with the California Floristic Region and its rapidly evolving vegetation types. *Hesperolinon* is a derivative of perennial Madrean groups of *Linum* that occur farther east (Rogers, 1975). The mediterranean climate of this region has evolved only in latest Tertiary time (Axelrod, 1973) and has never been in direct contact with any other region with a mediterranean climate, this explaining in part the very high endemism in California.

Despite this, genera such as *Arbutus, Cupressus, Erodium, Lavatera,* the evergreen white oaks, and the annual Inuloid Asteraceae of the western United States appear to be related directly to species of the Mediterranean region or the Near East. We have no reason to think that they or their immediate ancestors have ever been widely distributed in the northern temperate forests. Axelrod (1975) has termed such links "Madrean-Tethyan" (cf. table 5).

A number of additional groups are mainly or entirely associated with the California Floristic Province, insofar as we can interpret their ranges at present. In these examples (table 6), it is usually unclear whether the ultimate origin of the group lies with northern, semiarid middle-latitude, or tropical elements. Only a few of these genera are found in the Old World, and in these cases, other elements are involved: *Euphorbia* and *Papaver* may be examples. A few species of primarily North American groups, such as *Coreopsis, Mimulus, Montia,* and *Solidago,* also occur in the Old World, and some primarily western North American groups, such as *Claytonia, Dodecatheon,* and *Polemonium,* reach eastern Siberia.

In view of the progressive geographical restriction of sclerophyllous vegetation from the mid-Pliocene onward, it is not surprising that a number of "Californian" plants characteristic of Madrean vegetation occur isolated in central or southern Arizona (table 7). Until the Late Pliocene, chaparral species doubtless occurred here and there across the area of

TABLE 5

Madrean-Tethyan Elements in the Flora of California

Asteraceae-Inuleae [annual genera; their relatives are found in W. Eurasia-N. Africa, although the exact relationships remain to be determined. *Logfia* (*Filago* sensu Munz and Keck, 1959) is disjunct between W. North America and this region (Chrtek and Holub, 1963; Holub, 1976). If the single North American species of *Micropus* actually is to be referred to that genus or to *Bombycilaena* it would provide a second instance of such a pattern disjunction, but it might be more closely related to North American species (J. Holub, personal communication). *Stylocline*, with about 6 species, is confined to the S.W. U.S. and adjacent Mexico; the single Old World species that has been referred to this genus is now referred to *Cymbolaena* and not regarded as directly related. *Hesperevax* (*Evax* sensu Munz and Keck, 1959) is best regarded as an endemic of western North America (see also Wagenitz, 1969; J. Holub, personal communication). *Psilocarphus* (Cronquist, 1950) has 4 species in W. North America and 1 disjunct in W. South America. *Logfia* and *Stylocline* range out onto the deserts.]

Ericaceae-Arbuteae (*Arctostaphylos*, 50 species, all in Calif., but a few more widespread and 1 circumboreal; P. V. Wells, 1968 and personal communication; Roof, 1976; *Arbutus*, 20 species, scattered across Madrean-Tethyan region, Axelrod, 1975, 1 on Pacific Coast; *Comarostaphylis*, mainly Mexican, 20 species, 1 in chaparral of southern Calif. and N.W. Baja Calif.; the monotypic *Ornithostaphylos* endemic to the same region, Moran, 1973; another monotypic genus, *Xylococcus*, from coastal S. Calif. S. at least to Volcán las Tres Vírgenes in S.-cent. Baja Calif., Moran, personal communication.)

Pteridaceae (*Cheilanthes, Pellaea*; Howell, 1960)

Aesculus [Hippocastanaceae: 13, 8 in North America, 5 in Eurasia, 1 endemic in CFP. *Aesculus parryi* A. Gray, the least specialized species of the genus, is endemic in N.W. to cent. Baja Calif., and *A. californica* (Spach) Nutt. around the Central Valley of Calif.; the other 4 species of its section are Asian; 6 species representing 2 other sections occur in the cent. and E. United States, and one further section, with 2 species, occurs in W. Eurasia and in Japan; the only other genus of the family, the ditypic *Billia*, is tropical American and *Aesculus* certainly originated in North America.]

Antirrhinum (Scrophulariaceae: 12 species in Pacific North America, centering in CFP, where 8 are endemic, and about 20 in the W. Mediterranean, where a similar pattern of narrow endemism occurs in the Iberian Peninsula)

Aristolochia (Aristolochiaceae: 350, widespread, 1 endemic species in Calif.)

Astragalus (Fabaceae-Faboideae: 2000, of which 400 North America, 100 South America, 1 South Africa; in Calif. 94 species, 61 in CFP, 44 endemic, with 33 species in Calif. outside of limits of CFP. Although some species of *Astragalus*, of which *A. canadensis* L. is an example, seem to belong with the Arcto-Tertiary group, the vast majority in North America have evolved in Madrean vegetation, and many of these have entered Calif. only since mid-Pliocene time. As in *Vicia*, annual species have evolved both in the Old World and in the New; Barneby, 1964.)

Centaurium (Gentianaceae: 50, cosmopolitan except tropics and South Africa, with 8 species in Calif., 5 endemic to CFP; Broome, 1974, 1977)

Cercis (Fabaceae–Caesalpinioideae: 7, Mediterranean region to central China, warm temperate North America; 1 in Calif.)

Cicendia (*Microcala*; Gentianaceae, 2, 1 in W. Eurasia and N. Africa, the other in CFP and in Chile; Raven, 1963)

Comandra (Santalaceae: 2, 1 widespread in temperate North America, including Calif., and 1 in the Balkans and Romania; Piehl, 1965)

Convolvulus (Convolvulaceae: 250; *C. simulans* L. M. Parry, our only species, is an annual, endemic to the CFP but said by Perry, 1931, to be closely related to the Mediterranean *C. pentapetaloides* L.; these relationships should be investigated biosystematically, especially since there seems to be no other species of the genus in the New World)

Cupressus (Cupressaceae: 25-30 species, more than half in and near California, central Arizona to west Texas and adjacent Mexico, and central Mexico to Costa Rica; the others in southern Europe and western Asia; Wolf, 1948; Little, 1970)

TABLE 5 (continued)
Madrean-Tethyan Elements in the Flora of California

Datisca (Datiscaceae: 2, 1 endemic to CFP, the other from Crete and Turkey to the Himalayas; Davidson, 1973)

Erodium (Geraniaceae: 90, Mediterranean region to Cent. Asia, temperate Australia, S. tropical South America, 2 in Calif., 1 widespread in S.W. United States and adjacent Mexico, 1 endemic from Cent. Calif. to N. Baja Calif. Both belong to subsect. *Malacoides* which is common to the Old World.)

Galium (Rubiaceae: 400, cosmopolitan. The representatives of sect. *Lophogalium* K. Schuman that occur in W. North America include the dioecious *G. multiflorum* Kell. complex, with 16 species of W. North America, 7 in Calif., 2 endemic—primarily a group of the Great Basin, together with 14 polygamodioecious species, 6 in the CFP; Dempster and Ehrendorfer, 1965; Dempster, 1973. A group of 12 fleshy-fruited species treated by Dempster and Stebbins, 1968, is entirely endemic to the CFP. There are at least 6 other native species possibly more closely related to Eurasian groups and mentioned in table 4.)

Helianthemum (Cistaceae: 130, about 100 of W. Eurasia and N. Africa, the rest of North America, mostly S.E. United States and Mexico, sometimes regarded as a separate genus, *Crocanthemum*; 2 species recorded from Chile, but Muñoz (1966, p. 76), says in error. Relationships between Old and New World species would be most interesting to study.)

Lavatera (Malvaceae: 25, Eurasia, Australia, but with 4 species on islands off coast of Calif. and Baja Calif. Relationships badly in need of investigation but affinities definitely with European genera, unlike all other California Malvaceae except *Sidalcea*; D. M. Bates, personal communication.)

Loeflingia (Caryophyllaceae: 5 W. Mediterranean, 1 Great Plains to Calif. American species polytypic, with 4 subspecies, often cleistogamous; Barneby and Twisselmann, 1970)

Lotus (Fabaceae-Faboideae: 250, but all of the native species of the W. United States, about 50 in number, are probably best segregated as the endemic genus *Hosackia*; about two-thirds of them are found in Calif. The relationships of *Hosackia* to the Old World species of *Lotus* would be an interesting subject for investigation; Grant and Sidhu, 1967.)

Malvella [Malvaceae: 4, 1 in Europe and Middle East, 3 widespread in warmer areas of New World, 1 in Calif.; Fryxell, 1974. *Sida hereracea* (Dougl.) Torr. = *M. leporosa* (Ortega) Krapov.]

Platanus (Platanaceae: 10, with *P. orientalis* L. from S.E. Europe to W. Himalaya and *P. kerrii* Gagnep. in Indochina. *Platanus occidentalis* L. of E. North America is related to about 5 species of the Sierra Madre Oriental of Mexico, S. to Guatemala. *Platanus racemosa* Nutt. of the CFP S. to Baja Calif. and the closely related, possibly conspecific *P. wrightii* S. Wats. of Ariz. S. to N. Chihuahua and Sonora are directly related to *P. orientalis*, indicating Madrean-Tethyan, not northern, connections; Hsiao, 1973.)

Polygala (Polygalaceae: 600, cosmopolitan, 4 in Calif., 2 endemic; not a northern genus)

Prunus subg. *Amygdalus* (Rosaceae: 40, Mediterranean region to central China, with about 6 from S.E. Ore. to Texas, S. to central Mexico; the North American species sometimes referred to the segregate genus *Emplectocladus*. The 3 in Calif. mostly east of the Sierra Nevada, but *P. fasciculata* (Torr.) A. Gray var. *punctata* Jeps. is endemic to N. Santa Barbara and S. San Luis Obispo Co. on the coast.)

Prunus subg. *Laurocerasus* (Rosaceae: 75, tropical and temperate Asia and America, 1-2 species in tropical Africa and Madagascar, 1 w. Mediterranean region, 1 S.E. Europe to Caucasus. Two closely related taxa in the CFP, both S. to central Baja Calif.)

Psoralea (Fabaceae-Faboideae: 130, warmer parts of the Old World and well represented in North America; 8 species in Calif., of which 6 endemic to CFP)

Quercus (Fagaceae: 450. The evergreen white oaks and perhaps some other groups appear to show direct links between the Mediterranean region and North America; Axelrod, 1975.)

Trifolium (Fabaceae-Faboideae: 300, nearly cosmopolitan, with 40 species in CFP; Gillett, 1976. Although the genus may have migrated through the northern forests, annuals have evolved independently in the Mediterranean region and in the western United States.)

Triodanis (Campanulaceae: 8, with 2 widespread species in Calif., 1 in Mediterranean region, others in temperate North America)

TABLE 6
Taxa Primarily Associated with the California Floristic Province

Asteraceae-Helenieae (the endemic genera *Baeriopsis, Lembertia, Monolopia, Orochaenactis, Pseudobahia,* and *Venegasia*; 3 endemic and 2 nonendemic species of the North American genus *Helenium*; *Chaenactis,* radiating onto the deserts; as well as *Amblyopappus, Eriophyllum, Lasthenia* listed separately. The Helenieae are clearly a North American group and have radiated here.)

Asteraceae-Heliantheae (*Balsamorhiza; Coreopsis,* with 2 endemic sections including 8 species, 2 ranging onto the desert; *Eastwoodia,* monotypic endemic shrub of inner S. Coast Ranges; *Helianthella; Helianthus; Rudbeckia*; and *Wyethia*)

Asteraceae-Heliantheae, subtribe Madiinae (see p. 6)

Asteraceae-Lactuceae, subtribe Stephanomeriinae (most species of *Stephanomeria* [13/14: Gottlieb, 1972, 1973a] and *Malacothrix* [14/16] occur in Calif., but 11 of the latter and only 4 of the former are endemic to the CFP, where 1 of the 2 species of *Rafinesquia* also occurs. The small genera *Atrichoseris* and *Glyptopleura* occur in Calif. beyond the borders of the CFP, as do *Anisocoma* and *Calycoseris* which enter the CFP in the Sierra Juárez of Baja Calif.; there has been much radiation of the subtribe in semiarid vegetation types: Stebbins, 1953; Tomb, 1974.)

Brassicaceae (*Athysanus, Heterodraba, Thysanocarpus,* and *Tropidocarpum* are small genera more or less associated with mediterranean-type vegetation in Calif., and doubtless derived from other genera more widespread in more arid situations in western North America.)

Brassicaceae-Thelypodieae (10 genera, about 100 species, centered in Madrean vegetation in western North America, with the ditypic *Macropodium* in Siberia and temperate eastern Asia; Al-Shehbaz, 1973. For each of the genera in Calif. the total number of species, number in Calif., number in the CFP, and number endemic is given: *Caulanthus,* 15, 15, 11, 7; *Stanleya,* 6, 3, 1, 0; *Streptanthella,* 1, 1, 1, 0; *Streptanthus,* 30-35, 24, 22, 22; *Thelypodium,* 18, 8, 4, 1. Edaphic endemism in Calif. is very pronounced in *Streptanthus,* particularly with relation to serpentine. If these are in fact the least specialized genera of the family, it may have originated from Capparaceae-like ancestors in Madrean vegetation in W. North America.)

Hydrophyllaceae (see p. 51)

Limnanthaceae (2 genera and 10 species, 1 wide ranging in North America, 1 endemic to Vancouver Island, and the remainder near endemics of the CFP; Ornduff, 1971)

Onagraceae, tribes Epilobieae and Onagreae (see pp. 51-52)

Papaveraceae (11 genera and 23 species in Calif., 9 and 21, of which 18 are endemic, in the CFP; *Arctomecon* and *Canbya* are endemic just outside of the CFP but occur within the State. Three of the 4 subfamilies delineated by Ernst, 1962, occur in Calif. and Platystemonoideae are virtually endemic; Eschscholzioideae seem to have radiated from a subhumid climate—cf. *Dendromecon*—onto the desert, and are also endemic to the southwestern United States and adjacent Mexico.)

Polemoniaceae, tribe Gilieae (see p. 52)

Polygonaceae, subfamily Eriogonoideae (see pp. 53-54)

Portulacaceae (5 genera with 43 species, 16 endemic, in CFP; Heckard and Stebbins, 1974; Hinton, 1975, 1976a, 1976b; Miller, 1976. Represented in Eurasia only by *Montia fontana* L. sens. lat. and *Portulaca oleracea* L., together with 8 species of *Claytonia* in Siberia; thus the genera in and around Calif. are probably basically North American and probably ultimately of Southern Hemisphere origin.)

Saxifragaceae-Hydrangeoideae (*Carpenteria, Fendlerella, Jamesia, Philadelphus,* and *Whipplea* in Calif. The subfamily probably originated in Madrean vegetation in S.W. North America, and *Carpenteria,* the least specialized genus, as well as *Whipplea,* both monotypic, are endemic to the CFP; certain genera, like *Philadelphus,* have reached Eurasia and in some instances become speciose there; Stebbins, 1974, p. 182-183.)

Acanthomintha (Menthaceae: 3, endemic to Calif. and N. Baja Calif.)

Adenostoma (Rosaceae-Rosoideae: 2, CFP S. to Cedros I. and Sierra San Borja; related to genera such as *Purshia,* of arid interior North America, barely entering the CFP, but now confined to summer-dry regions)

Ageratina (Asteraceae, tribe Eupatorieae: 230, E. United States to W. South America, 2 in Calif., 1 in CFP, both widespread)

TABLE 6 (continued)
Taxa Primarily Associated with the California Floristic Province

Amblyopappus (Astereae, Helenieae or Senecioneae; Skvarla and Turner, 1966: 1, coastal S. Calif. and adjacent Baja Calif.; Peru to cent. Chile)

Amorpha (Fabaceae-Faboideae: 15, 2 in Calif., 1 endemic, 1 additional endemic in N.W. Baja Calif.)

Asclepias (Asclepiadaceae: 120, N. and S. America, 13 in Calif., 8 in CFP, 3 endemic; seems to have radiated from Madrean area in North America; Woodson, 1954)

Aspidotis (Pteridaceae: 3, Calif., Mexico, tropical Africa; Pichi-Sermolli, 1950)

Benitoa (Asteraceae, Astereae: 1, endemic, inner S. Coast Ranges, derived from *Haplopappus* complex of the S.W. deserts)

Blennosperma (Astereae, Senecioneae: 3, 2 in Calif., 1 in cent. Chile; closely related to *Crocidium*; Ornduff, 1963, 1964; Skvarla and Turner, 1966; Ornduff et al., 1973)

Bloomeria (Amaryllidaceae: 2, endemic; Hoover, 1955)

Brodiaea (Amaryllidaceae: 15, all endemic except for 2 that range north beyond the CFP; Niehaus, 1971)

Ceanothus (Rhamnaceae: 53, with 44 in Calif., 37 endemic to CFP; widespread in Madrean vegetation in Tertiary time, but now mainly a rapidly evolving complex of CFP. See pp. 78-79)

Cercocarpus (Rosaceae-Rosoideae: 8, 5 in Calif., with 3, all endemic, in the CFP; others in analogous shrubland vegetation of Mexico and adjacent United States derived from Madro-Tertiary Geoflora)

Chamaebatia (Rosaceae-Rosoideae: 2, endemic to CFP)

Chlorogalum (Liliaceae: 5, endemic and just reaching Baja Calif. and S.W. Ore. Regarded by Hutchinson (1973) as one of many genera of the worldwide tribe Liliaceae-Asphodeleae, but by Huber (1969), probably with better reason, as an isolated genus forming its own tribe Chlorogaleae of Hyacinthaceae. In any case, a long history in Madrean vegetation is indicated, and many aspects of the group would amply repay further study.)

Cneoridium (Rutaceae: 1, endemic, Orange Co., Calif., to Sierra San Borja, cent. Baja Calif.)

Collinsia (Scrophulariaceae: 17, 15 in Calif. and 2 in cent. and E. United States; the ditypic genus *Tonella* is closely related)

Corethrogyne (Asteraceae, Astereae: 3, endemic, closely related to *Lessingia*; 12, nearly endemic; presumably derived from *Machaeranthera*-like ancestors of the S.W. deserts and evolving in Calif. since Middle Pliocene time; Spence, 1963)

Crocidium (Astereae, Senecioneae: 1, B.C. south to cent. Calif.; Ornduff et al., 1973)

Dichelostemma (Amaryllidaceae: 6 species, 4 endemic, 1 north to Wash., 1 south to Sonoran Desert; Hoover, 1940; Keator, 1968; Lenz, 1974)

Downingia (Campanulaceae: 14, 1 in South America, 1 common but restricted to CFP in North America, 5 endemic to CFP, 4 occurring within it but ranging outside, and 2 of the N.W. Great Basin)

Dudleya (Crassulaceae: 40, more than half in Calif. and these almost all endemic; many in deserts and semiarid mountains southward, especially in Baja Calif.)

Eremocarpus (Euphorbiaceae: 1, Calif. to Wash.)

Eriophyllum (Asteraceae, Helenieae: 13, all in Calif., but 1 of the 6 perennials ranging widely in the W. United States and 4 of the 7 annuals outside of the CFP on the deserts)

Fremontodendron (Sterculiaceae: 2, Calif. to N. Baja Calif., cent. Ariz.; a very isolated genus of uncertain affinities)

Githopsis (Campanulaceae: 6, nearly endemic to CFP, but 1 north to Wash.)

Hesperolinon (Linaceae: 12, all in Calif. but one ranging widely; Sharsmith, 1961. Ultimately Madrean and derived from *Linum*; Rogers, 1975)

Heterocodon (Campanulaceae: 1, W. North America)

Lasthenia (Asteraceae, Helenieae: 16, 1 in Chile, 15 in Calif., 3 ranging beyond the borders of the CFP; Ornduff, 1966)

Legenere (Campanulaceae: 2, 1 endemic in Sacramento Valley, the other in Chile and Argentina)

Lyonothamnus (Rosaceae-Spiraeoideae: 1 or 2, endemic to 4 islands off coast of S. California; a primitive genus of the subfamily)

Malacothamnus (Malvaceae: 11, endemic to Calif. but with 1 sp. ranging onto the deserts; related to *Iliamna*, of more mesic habitats, and perhaps with it derived from the Mexican, Central American, and West Indian *Phrymosia*; Bates, 1963 and personal communication; Bates and Blanchard, 1970)

TABLE 6 (continued)
Taxa Primarily Associated with the California Floristic Province

Malosma (Anacardiaceae: 1, endemic to S. Calif. and Baja Calif., S. to Cape Region)

Marah (Cucurbitaceae: 7, 6 in Calif., 1 in Ariz. and N.M., 5 in CFP, of which 3 endemic)

Mimulus (Scrophulariaceae: 100, a few widespread but the great majority in and near Calif., where there are 77 species including the nearly endemic woody sect. *Diplacus*, one species of which extends to Sierra San Borja of cent. Baja Calif., another is endemic on Cedros I.)

Monardella (Menthaceae: 32, all restricted to the CFP except for a few that are more widespread, 2 that are endemic to the Mohave Desert of Calif., 1 to cent. and S. Baja Calif., and 1 endemic to cent. Ariz.; Epling, 1925; Hardham, 1966a, 1966b)

Muilla (Amaryllidaceae: 4, 2 endemic to CFP, 1 of Mohave Desert, 1 of N.W. Great Basin; Hoover, 1955)

Nemacladus (Campanulaceae: 13, 9 endemic to CFP, 3 in nearby deserts, and 1 in both areas)

Neostapfia (Poaceae: 1, endemic to Central Valley of Calif.; Crampton, 1959)

Orcuttia (Poaceae: 7, 6 endemic, 1 on Magdalena Plain of S. Baja Calif.; Crampton, 1959)

Parishella (Campanulaceae: 1, S. Calif. to deserts)

Parvisedum (Crassulaceae: 4, endemic. An annual group evidently derived from the circumboreal genus *Sedum;* Clausen, 1975)

Pentachaeta (Asteraceae, Astereae: 6, endemic to Calif. and northernmost Baja Calif.; Van Horn, 1973; Nelson and Van Horn, 1976)

Pickeringia (Fabaceae-Faboideae: 1, endemic to CFP)

Plectritis (Valerianaceae: 4, 3 of W. North America, 1 of S. Cent. Chile and adjacent Argentina; Raven, 1963)

Pogogyne (Menthaceae: 5, endemic to CFP)

Porterella (Campanulaceae: 1, W. United States including N. Sierra Nevada of Calif.)

Ptelea (Rutaceae: 3, 1 widespread in temperate North America, the other 2 endemic in CFP, 1 around the Central Valley of Calif., the other in N.W. Baja Calif.; Bailey, 1962)

Rigiopappus (Asteraceae, Astereae: 1, W. United States; very closely related to the monotypic, endemic *Tracyina* of N. Calif.; Ornduff and Bohm, 1975)

Salvia sect. *Audibertia* (Menthaceae: 19, centering in and around Calif., including 17 shrubby species and 2 annuals; 11 species occur in the CFP, of which 9 are endemic; the other 2 are widespread. Of the remaining 8 species, 1 is restricted to cent. Baja Calif., the other 7 are endemic around the borders of the CFP, with *S. eremostachya* Jeps. extending S. to the Sierra San Borja of cent. Baja Calif.; Epling, 1938; Epling et al., 1962.)

Sclerolinon (Linaceae: 1, endemic to Pacific States)

Sidalcea (Malvaceae: 22, W. North America, 18 in Calif., including 1 derived group of 5 annual species, all endemic)

Trichostema (Menthaceae: 16, with 9 of the 10 species in Calif. endemic to the CFP; Lewis, 1945)

Triteleia (Amaryllidaceae: 14. Sect. *Calliprora*, with 4 species, is endemic to the CFP, Lenz, 1975; sect. *Triteleia* has 5 species, 4 endemic and 1 ranging to B.C.; and sect. *Hesperocordum* also has 5 species, 2 endemic, 2 ranging north out of the CFP, and 1 in central Ariz.; Hoover, 1941. The related monotypic genus *Triteleiopsis* is endemic to the Sonoran Desert.)

Umbellularia (Lauraceae: 1, endemic. Possibly related to *Laurus* and if so reflecting Madrean-Tethyan connections.)

the present Mohave Desert, as indicated by Clements (1936), and supplemented by the recent report of Leskinen (1975). Another example that documents formerly more extensive distributions across the desert is provided by the occurrence of a single white oak hybrid of which one parent is *Quercus lobata* Née, in the Little San Bernardino Mountains; the nearest individuals of this species now grow about 240 km to the west (Tucker, 1968). Most of the desert area of Nevada, California and western Arizona was occupied by xeric conifer woodland up to about 10,000 years ago, and disjunct taxa that we see now are only the remnants, as inferred by Clements (1936).

TABLE 7
Some Taxa which Include "Californian" Elements Isolated in Central and Southern Arizona

Amsinckia	*Pityrogramma*
Arbutus	*Platanus*
Athysanus	*Platystemon*
Calandrinia	*Quercus*
Ceanothus	*Rhamnus*
Cercis	*Scrophularia*
Clarkia	*Thysanocarpus*
Cupressus	*Triteleia*
Dudleya	
Eriodictyon	*Arctostaphylos pringlei* Parry
Fremontodendron	*Camissonia intermedia* Raven
Lasthenia	*Cercocarpus betuloides* Nutt.
Lessingia	*Lilium parryi* S. Wats.
Marah	*Lonicera interrupta* Benth.
Microseris	*Rhus ovata* S. Wats.
Monardella	*Salix laevigata* Bebb
Perideridia	*Solanum umbelliferum* Eschs. sens. lat.
Pholistoma	*Yucca whipplei* Torr.-*Y. newberryi* McKelvey

In some instances, taxa associated with Madrean vegetation types are difficult to distinguish from the taxa of Arcto-Tertiary affinities. For example, *Arceuthobium* (Viscaceae), with 28 species, is considered to have Arcto-Tertiary affinities because four species are found in the Old World, suggesting that the genus may have had a northern origin, even though most of the New World species are parasitic on conifers in Madrean vegetation (Hawksworth and Wiens, 1972). *Perideridia* (Apiaceae), with 14 species, like other apioid members of the family, appears to have had a northern origin. One species occurs in Japan and Korea (Chuang and Constance, 1969), despite the fact that the American species barely reach southern Vancouver Island and a majority are endemic to the California Floristic Province.

Warm Temperate and Desert Elements

Although the deserts that border the California Floristic Province have only recently attained regional extent (see pp. 43-46), many of the taxa in them were in existence in the Miocene and earlier in communities resembling those that now border the regional deserts. The continued trend toward spreading drought, as in the Xerothermic periods of the Quaternary, allowed many taxa that now are primarily associated with deserts to invade the drier parts of the California Floristic Province. They have become especially prominent in the San Joaquin Valley and up into the inner South Coast Ranges, as well as in the interior valleys of southern California. These taxa are enumerated in table 8. In this group we have also included certain taxa that are widespread in the warmer parts of North America but not particularly associated with the deserts.

Approximately 188 genera seem to have radiated into the California Floristic Province from semiarid or desert areas or have come into the region from warmer, subtropical areas. All of the species of these genera that are known to occur in the state have been listed in table 8. They include about 756 species in California, or about 15% of the flora of the State, and 604 species in the California Floristic Province or 13.5% of the total.

TABLE 8

Primarily Warm Temperate or Desert Taxa in the California Floristic Province

Asteraceae, subtribe Ambrosiinae (*Ambrosia, Dicoria, Hymenoclea, Iva,* and *Xanthium* have a total of 18 species in Calif., including 1 of *Iva* endemic to the S. coastal part of the CFP; the subtribe clearly originated in the arid regions of W. North America; Payne, 1964.)

Asteraceae, tribe Heliantheae (*Encelia, Geraea, Verbesina,* and *Viguiera,* with a total of a dozen species in Calif. and 2 endemics in the CFP, represent a complex of genera that has its main center of evolution in the deserts of the S.W. United States and adjacent Mexico.)

Crossosomataceae (Three genera: *Crossosoma,* with 2 species, has 1 in cent. Ariz., S.E. Calif., N.W. to cent. Baja Calif., and 1 on islands off southern Calif. and Baja Calif. The monotypic *Apacheria* is endemic in S. Ariz.; Mason, 1975. *Forsellesia* has 8 species in the W. U.S., 3 in Calif., 1 reaching the Trinity R. Canyon, Trinity Co.; R. F. Thorne, personal communication, 1977.)

Cyperaceae (For the following widespread genera of mostly moist temperate to subtropical situations are given the total species, species in Calif., species in CFP, and endemics: *Cladium* [40, 1, 1, 0], *Cyperus* [600, 12, 12, 0], *Eleocharis* [200, 9, 9, 0], *Fimbristylis* [300, 3, 2, 0], *Hemicarpha* [6, 2, 2, 0], *Scirpus* [200, 14, 14, 3]. The 3 endemics and a few other species in *Scirpus* seem to have N. affinities, as do a few species of *Eleocharis.*)

Lennoaceae (*Lennoa* has 3 species in Cent. Mexico; *Ammobroma* 2 from the Sonoran Desert, including Calif., to Sinaloa; and *Pholisma* 1 or 2 on deserts of S. Calif. and Baja Calif. and N. along the coastal sands to San Luis Obispo Co.)

Lythraceae (The family includes *Ammania, Lythrum,* and *Rotala,* widespread genera that constitute its total representation, including 5 species, in Calif., with 1 of *Lythrum* a near endemic.)

Poaceae (The following genera are widespread in semiarid to tropical regions of the New World and sometimes also the Old. For each is listed total species, species in Calif., species in the CFP, and endemics: *Andropogon* [113, 2, 2, 0], *Aristida* [330, 12, 12, 0], *Blepharoneuron* [1, 0, 1, 0], *Bothriochloa* [30, 1, 1, 0], *Bouteloua* [40, 9, 4, 0], *Cenchrus* [25, 2, 2, 0], *Chloris* [40, 1, 1, 0], *Dichanthelium* [25, 3, 2, 0; Gould, 1974; Spellenberg, 1975], *Digitaria* [380, 0, 1, 0], *Distichlis* [12, 1, 1, 0], *Eragrostis* [300, 9, 11, 0], *Eriochloa* [20, 2, 1, 0], *Imperata* [10, 1, 1, 0], *Leptochloa* [27, 3, 1, 0], *Lycurus* [3, 0, 1, 0], *Monanthochloë* [1, 1, 1, 0], *Muhlenbergia* [100, 16, 18, 2], *Panicum* [350, 3, 2, 0; Gould, 1974; Spellenberg, 1975], *Paspalum* [250, 2, 2, 0], *Setaria* [140, 1, 1, 0], *Sporobolus* [150, 4, 1, 0].)

Polygonaceae, see pp. 53-54.

Scrophulariaceae (*Bacopa, Gratiola, Limosella, Lindernia,* and *Stemodia* are widespread genera of warm to temperate areas, aquatic or subaquatic, represented by a total of 11 species in Calif., with 1 each of *Bacopa* and *Gratiola* endemic)

Abronia (Nyctaginaceae: 17, W. North America, 8 in Calif. and CFP, deserts to the coastal sands, 2 endemic, 1 on the coast, 1 in high sandy meadows of the S. Sierra Nevada; Wilson, 1972; Galloway, 1976)

Abutilon (Malvaceae: 100, tropical and subtropical, 3 in deserts of Calif., 1 reaching Sierra Juárez, Baja Calif.; 1 additional species in N.W. Baja Calif.)

Acacia (Fabaceae-Mimosoideae: 800, tropical and subtropical, 1 in deserts of Calif., 1 probably native around San Diego and S. along W. coast of Baja Calif.)

Acalypha (Euphorbiaceae: 450, tropics and subtropics, 1 endemic in San Diego Co. and Baja Calif.)

Achyronychia (Caryophyllaceae: 2, S.W. U.S. and adjacent Mexico, 1 on deserts of Calif. and extending N. to Santa Maria, S. of San Quintín, Baja Calif.)

Acourtia (Astereae-Mutisieae: 41, *Perezia* in part, S.W. United States to Central America, 1 in CFP, endemic; Reveal and King, 1973)

Adolphia (Rhamnaceae: 2, 1 of the N.W. quarter of Mexico, the other endemic to cismontane San Diego Co., Calif., and N.W. Baja Calif.; Shreve and Wiggins, 1964, p. 867-868)

Agave (Agavaceae: 300, semiarid to tropical regions of the Americas; 3 in Calif., 2 on deserts, 2 endemic to CFP, 1 of them in Baja Calif. only)

Allenrolfea (Chenopodiaceae: 4, arid North and South America, 1 to San Joaquin Valley)

Amaranthus (Amaranthaceae: 60, tropical and temperate, 5 in CFP, none endemic, 2 others in deserts)

Amauria (Asteraceae-Helenieae: 3, N.W. Mexico, 1 N. to vicinity of San Quintín, Baja Calif.)

Anemopsis (Saururaceae: 1, Sacramento Valley to Baja Calif., E. to Texas and adjacent Mexico, in semiarid and often alkaline places)

TABLE 8 (continued)
Primarily Warm Temperate or Desert Taxa in the California Floristic Province

Aphanisma (Chenopodiaceae: 1, endemic to coastal S. Calif. and adjacent Baja Calif.)

Arthrocnemum (Chenopodiaceae: 20, coasts Mediterranean region to Australia, warm North America, 1 to San Joaquin Valley and San Francisco Bay)

Atriplex (Chenopodiaceae: 200, temperate and subtropical, cosmopolitan, 27 in Calif., 21 CFP, 11 endemic, 4 in the Central Valley)

Baccharis (Asteraceae-Astereae: 400, Americas, mainly South America, 9 in Calif., 7 in CFP, 2 endemic, both coastal, and 1 near endemic coastal species)

Bahia (Asteraceae-Helenieae: 15, 2 widespread species in Calif., 1 to San Bernardino Mts.)

Baileya (Asteraceae-Helenieae: 3, S.W. U.S. and N.W. Mexico, all on deserts of Calif., 1 reaching CFP in N.W. Baja Calif.)

Batis (Bataceae: 2, Pacific, 1 in warm regions of Americas including S. Calif.)

Bebbia (Asteraceae-Heliantheae: 2, 1 widespread in Sonoran and Mohave deserts and reaching cismontane S. Calif.; Whalen, 1977)

Bergerocactus (Cactaceae: 1, endemic to coastal S. Calif. and adjacent Baja Calif.)

Bernardia (Euphorbiaceae: 50, tropical and subtropical America, 1 in Sonoran Desert of Calif. and entering CFP in Sierra Juárez, Baja Calif.)

Bidens (Asteraceae-Heliantheae: 230, cosmopolitan, wet places, 2 in Calif. and 2 additional species in N.W. Baja Calif.)

Blepharoneuron (Poaceae: 1, S.W. U.S. and adjacent Mexico, including N.W. Baja Calif.)

Boerhavia (Nyctaginaceae: 40, tropical and subtropical, 6 in Calif., 1 to cismontane S. Calif.)

Brahea (Arecaceae: 15, N.W. Mexico to Costa Rica, 1 N. to Sierra San Pedro Mártir, 1 endemic on Guadalupe I.; including *Erythea*)

Brickellia (Asteraceae-Eupatorieae: 100, New World, 14 in Calif., 5 in CFP, 2 endemic)

Buddleja (Buddlejaceae: 100, tropical and subtropical, 1 in Inyo region of Calif., another reaching CFP at Hamilton Ranch, on Río Santo Domingo, N.W. Baja Calif.)

Calycoseris (Asteraceae-Lactuceae: 2, deserts of S.W. U.S. and N.W. Mexico, both in Calif., 1 reaching CFP in Sierra Juárez and S.W. San Pedro Mártir, Baja Calif.)

Cephalanthus (Rubiaceae: 17, warmer parts of Americas and Asia, 1 widespread species in Calif.)

Ceratoides (Chenopodiaceae: 8, 1 semiarid W. North America to San Joaquin Valley, the others Eurasia to N. Africa; *Eurotia*, Howell, 1971)

Chamaesyce (Euphorbiaceae: 250, cosmopolitan, with 12 species, including 2 endemic–1 in the Central Valley, 1 in N.W. Baja Calif., in the CFP and 11 additional species in Calif. but only on the deserts; Webster, 1967; R. Moran, personal communication)

Chenopodium (Chenopodiaceae: 100-150, cosmopolitan, 13 Calif., 13 CFP, 1 endemic)

Chrysothamnus (Asteraceae-Astereae: 13, arid W. North America, 9 in Calif., 2 reaching the Sierra Nevada crest and 1 the inner S. Coast Ranges)

Cleome (Capparaceae, including *Isomeris*: 150, tropics and subtropics, 5 in Calif., 1 reaching N.E. borders of CFP, another from deserts to cismontane S. Calif.; *C. jonesii* (J. F. Macbr.) Tidestr. reaching CFP in N.W. Baja Calif.)

Condalia (Rhamnaceae: 18, warm, N. and S. America, 1 reaching CFP near San Quintín, Baja Calif.)

Conyza (Asteraceae-Astereae: 60, temperate and subtropical cosmopolitan; 2 widespread species in Calif.)

Cressa (Convolvulaceae: 5, warm regions, widespread, 1 reaching Calif.)

Croton (Euphorbiaceae: 750, tropics and subtropics, 2 in Calif., 1 reaching sandy places in Central Valley and near the coast N. to San Francisco and Antioch)

Cucurbita (Cucurbitaceae: 20, Americas, 3 widespread species in Calif. and all reaching the CFP; Bemis and Whitaker, 1969)

Dalea (Fabaceae-Faboideae: 250, W. Hemisphere, 10 on deserts of Calif., 1 additional species reaching CFP in Sierra Juárez, Baja Calif.)

Digitaria (Poaceae: 380, tropical and subtropical, 1 to N.W. Baja Calif.)

Dracocephalum (Menthaceae: 45, Eurasia, 1 widespread N. America and N.W. Baja Calif.)

Drymaria (Caryophyllaceae: 45, mostly W. Hemisphere, 2 to N.W. Baja Calif.)

Dyssodia (Asteraceae-Helenieae: 40, S.W. U.S. and Mexico, 3 on deserts of Calif., 1 of them reaching CFP in Sierra Juárez, Baja Calif.)

TABLE 8 (continued)
Primarily Warm Temperate or Desert Taxa in the California Floristic Province

Echinocereus (Cactaceae: 30, S.W. U.S. and Mexico, 2 in deserts of Calif., 2 others in N.W. Baja Calif.)

Ephedra (Ephedraceae: 40, with 7 in Calif., but only *E. californica* S. Wats. enters the CFP, in San Diego Co. and the inner South Coast Ranges)

Eremalche (Malvaceae: 4, 2 endemic in and near San Joaquin Valley, 2 in Mohave and Sonoran deserts, one of them reaching CFP in N.W. Baja Calif.)

Eucnide (Loasaceae: 11, S.W. U.S. to Guatemala, 2 in deserts of Calif., 1 additional species reaching CFP in Sierra San Pedro Mártir, Baja Calif.; Thompson and Ernst, 1967)

Euphorbia subg. *Agaloma* (Euphorbiaceae: 100, American, mainly tropical and subtropical, with 1 endemic to S. Calif., Baja Calif., and N.W. Sonora)

Eustoma (Gentianaceae: 3, S. U.S. and Mexico, 1 to S. Calif.)

Evolvulus (Convolvulaceae: 100, tropical and subtropical, 1 to N.W. Baja Calif.)

Ferocactus (Cactaceae: 35, S.W. U.S. and Mexico, 1 endemic in S.W. San Diego Co. and adjacent Baja Calif., another reaching N.W. Baja Calif., a third in Calif. deserts, reaching CFP in N.W. Baja Calif.)

Forestiera (Oleaceae: 15, 1 from S.W. U.S. and adjacent Mexico to deserts and inner Coast Ranges of Calif.)

Fouquieria (Fouquieriaceae: 11, S.W. U.S. and Mexico, 1 on deserts of Calif., reaching CFP in foothills of Sierra San Pedro Mártir, Baja Calif.; Henrickson, 1972)

Grindelia (Asteraceae-Astereae: 50, W. North and South America, arid regions, 11 in Calif., all found in and 9 of them endemic to CFP)

Haplopappus (Asteraceae-Astereae: 150, W. North and South America, mostly arid regions, a complex genus that will be subdivided; 38 species occur in Calif., 37 in the CFP, and 21 are endemic to it, 7 in N.W. Baja Calif. only. In subg. *Ericameria*, there are about 16 species, 14 in the CFP and 12 of them endemic; the group appears to parallel *Diplacus* and *Keckiella* as a woody offshoot in the CFP from a more widespread group.)

Harfordia (Polygonaceae: 2, Baja Calif., N. to San Antonio del Mar in CFP, and Cedros Island)

Hedeoma (Menthaceae: 30, W. U.S. and Mexico, 1 in deserts of Calif., 1 endemic in Sierra San Pedro Mártir, Baja Calif.)

Heliotropium (Boraginaceae: 250 cosmopolitan species, mainly tropical, 2 in Calif., 1 throughout CFP)

Hibiscus (Malvaceae: 300, tropical and subtropical, 2 in Calif., 1 widespread American species in Central Valley, 1 in Sonoran Desert)

Hoffmanseggia (Fabaceae-Caesalpinioideae: 45, America and Africa, 2 on deserts of Calif., 1 extending into CFP in Baja Calif. and possibly as an adventive in S. Calif. also)

Hulsea (Asteraceae-Helenieae: 7, arid ranges of S.W. U.S., in Calif. with 2 endemic in CFP and 1 extending S. to N. Cent. Baja Calif.; Wilken, 1975)

Hymenopappus (Asteraceae-Anthemideae: 10, 1 widespread species to mts. of S. Calif. and Sierra San Pedro Mártir, N.W. Baja Calif.)

Hymenoxys (Asteraceae-Helenieae: 20, 4 in Calif., 1 to N.E. borders of CFP)

Hyptis (Menthaceae: 400, tropical and subtropical America, 1 on deserts of Calif. and in CFP at Punta Banda, Baja Calif.)

Ipomoea (Convolvulaceae: 500, tropical and subtropical, 1 widespread species to N.W. Baja Calif. in CFP)

Justicia (Acanthaceae: 450, tropical and subtropical America, 1 on deserts of Calif. to cismontane San Diego Co. and common in CFP of N.W. Baja Calif.; including *Beloperone*, Gibson, 1972)

Kochia (Chenopodiaceae: 90, mainly Old World, 2 Calif., 1 reaching San Joaquin Valley)

Lepidium (Brassicaceae: 150, cosmopolitan, 14 in Calif., 11 in CFP, 3 endemic, 2 in alkaline flats and valley floors, 1 in S. Coast Ranges)

Lepidospartum (Asteraceae-Senecioneae: 2, N. Sonoran Desert N. to Inyo Co., 1 in inner S. Coast Ranges to Santa Clara Co.)

Lesquerella (Brassicaceae: 40, 3 in Calif., 2 reaching borders of CFP, 1 endemic in Sierra San Pedro Mártir, N.W. Baja Calif., mostly arid interior W. North America)

Lobelia (Campanulaceae: 200-300, cosmopolitan, 2 in CFP, 1 nearly endemic but reaching desert canyons along E. side of Sierra San Pedro Mártir, N.W. Baja Calif.)

TABLE 8 (continued)
Primarily Warm Temperate or Desert Taxa in the California Floristic Province

Lophocereus (Cactaceae: 2, S. Ariz. to N.W. Mexico, 1 to N.W. Baja Calif. in CFP; Lindsay, 1963)

Lycium (Solanaceae: 80-90, temperate and subtropical, especially South America, 9 in Calif., 6 in CFP, 3 endemic)

Lycurus (Poaceae: 3, tropical and subtropical America, 1 to N.W. Baja Calif.)

Machaeranthera (Asteraceae-Astereae: 25, semiarid W. North America, 5 in Calif., 3 in and 1 endemic to CFP)

Machaerocereus (Cactaceae: 2, Baja Calif., W. Sonora, 1 to 8 km N. of Ensenada, Baja Calif.)

Mammillaria (Cactaceae: 100, 3 in Calif., 1 in San Diego Co., and adjacent Baja Calif., where 2 additional nonendemic species occur; R. Moran, personal communication)

Menodora (Oleaceae: 17, America and Africa, 3 in deserts of Calif., 2 of them reaching the CFP in Sierra Juárez, Baja Calif.)

Mentzelia (Loasaceae: 100, semiarid regions of North and South America, 28 in Calif., 10 in CFP, 5 endemic, 7 of the remaining 18 endemic on deserts of Calif.)

Mirabilis (Nyctaginaceae: 60, warmer parts of Americas, 6 Calif., 2 of 3 in CFP endemic; including *Hermidium*, Pilz, 1974)

Monolepis (Chenopodiaceae: 6, Asia, North America, Patagonia, 3 in Calif., 2 in CFP)

Myrtillocactus (Cactaceae: 4, Mexico and Guatemala, 1 to 13 km S. of Ensenada, Baja Calif.; R. Moran, personal communication)

Nitrophila (Chenopodiaceae: 8, semiarid regions of W. Hemisphere, 2 Calif., 1 in interior valleys of CFP)

Nolina (Agavaceae: 30, S.W. U.S., Mexico; 3 in Calif., 1 endemic to, and 2 entering CFP, 1 in Baja Calif. only)

Oligomeris (Resedaceae: 8, 1 common to Mediterranean region and arid W. North America, where perhaps introduced)

Opuntia (Cactaceae: 300, S. Canada to Straits of Magellan, 19 in Calif., 8 in CFP of which 3 endemic in coastal S. Calif. and adjacent Baja Calif.)

Oxybaphus (Nyctaginaceae: 25, America, 3 in deserts of Calif., 1 of them reaching the CFP in Sierra Juárez, Baja Calif.)

Pachycereus (Cactaceae: 5, N.W. Mexico, 1 to N.W. Baja Calif.)

Pectis (Asteraceae-Helenieae: 70, America, 1 in deserts of Calif. and to CFP in foothills of Sierra San Pedro Mártir, Baja Calif.)

Pennellia (Brassicaceae: 5, S.W. U.S. to South America, reaching Sierra San Pedro Mártir, ca. 2500 m; R. Moran, personal communication)

Pericome (Asteraceae-Helenieae: 4, S.W. U.S., adjacent Mexico, 1 widespread species reaching the Sierra Nevada)

Perityle (Asteraceae-Helenieae: 25, with 1 widespread species, mainly of the Sonoran Desert, reaching the coast of S. Calif. and Baja Calif. and 1 endemic on Guadalupe I.)

Petalonyx (Loasaceae: 5, S.W. U.S. and N.W. Mexico, 3 on deserts of Calif., 2 reaching southern borders of CFP in Baja Calif.; Davis and Thompson, 1967)

Phoenicaulis (Brassicaceae: 2, N.W. and W. Great Basin, but both in high Sierra Nevada)

Phoradendron (Viscaceae: 190, North and South America, mainly tropical with 4 species in CFP but none endemic; Wiens, 1964)

Physalis (Solanaceae: 100, cosmopolitan, mostly America, 7 in Calif., 4 in CFP, 1 endemic)

Plantago (Plantaginaceae: 265, cosmopolitan, 9 in CFP, all widespread)

Pluchea (Asteraceae-Inuleae: 50, tropics and subtropics, wet places, 2 widespread species in Calif.)

Porophyllum (Asteraceae-Helenieae: 30, 1 widespread species to coast of S. Calif.)

Portulaca (Portulacaceae: 100, tropical and subtropical, 1 in deserts of Calif., another reaching CFP in Sierra Juárez, Baja Calif.)

Proboscidea (Martyniaceae: 9, warm regions of the Americas, 2 in Calif. and 1 in CFP)

Prosopis (Fabaceae-Mimosoideae: 35, tropical and subtropics, 2 in deserts of Calif., extending to CFP in N.W. Baja Calif., 1 into San Joaquin Valley)

Rhus subg. *Schmaltzia* (Anacardiaceae: 50, warm-temperate North America, 3 in CFP; Barkley, 1937)

TABLE 8 (continued)
Primarily Warm Temperate or Desert Taxa in the California Floristic Province

Salicornia (Chenopodiaceae: 35, widespread on seacoasts and inland in alkaline places, 3 in Calif.)

Sanvitalia (Asteraceae-Heliantheae: 7, 1 on deserts of Calif., reaching CFP in Sierra Juárez, Baja Calif.; Torres, 1964)

Sarcostemma (Asclepiadaceae: 50, semiarid and subtropical habitats in North and South America, 1 reaching cismontane S. Calif., and in tropical Asia and Africa; Holm, 1950 and personal communication)

Sibara (Brassicaceae: 11, 4 in Calif., 2 endemic in and around Death Valley, 1 endemic on Santa Cruz I., 1 scattered and to E. U.S., 1 additional species to N.W. Baja Calif., 3 others farther S. in Baja Calif., the remaining 3 Texas to Mexico; Rollins, 1947)

Simmondsia (Buxaceae: 1, Sonoran Desert to near coast in San Diego Co.; very isolated and peculiar genus)

Sisyrinchium (Iridaceae: 100, Americas, 1 to Ireland; 7 in Calif., 1 endemic, 1 in Great Basin only; Mosquin, 1970; Henderson, 1976)

Solanum (Solanaceae: 1700, tropical and temperate; *S. douglasii* Dunal, a widespread species of *S. nigrum* L. group, common in Calif.; *S. triflorum* Nutt. just reaching N.E. Calif.; *S. palmeri* Vasey & Rose endemic to and *S. hindsianum* Benth. reaching N.W. Baja Calif.; and the polymorphic *S. umbelliferum* Eschs. group centering in the CFP but disjunct to Ariz.; W. F. Hinton, personal communication)

Sphaeralcea (Malvaceae: 60, 9 in Calif. and 4 reaching margins of CFP in Calif., up to several well within CFP in N.W. Baja Calif. but the specific limits poorly understood; deserts of S.W. U.S. and Mexico)

Stillingia (Euphorbiaceae: 30, tropics and subtropics, mainly New World, 1 reaching chaparral of S. Calif.)

Suaeda (Chenopodiaceae: 110, cosmopolitan, 5 Calif., 1 endemic along the coast)

Syntrichopappus (Asteraceae-Helenieae: 2, Mohave Desert, with 1 endemic to desert slopes of San Bernardino and San Gabriel Mts.)

Talinum (Portulacaceae: 50, widespread, 1 endemic, Guadalupe I.)

Tetracoccus (Euphorbiaceae: 4, 3 in Calif., 1 endemic in cismontane S. Calif. and N.W. Baja Calif.)

Tetradymia (Asteraceae-Senecioneae: 10, 8 in Calif., *T. comosa* A. Gray endemic in S. CFP, the only species in CFP; Strother, 1974)

Thamnosma (Rutaceae: 2, S.W. U.S. and N.W. Mexico, 1 in deserts of Calif., reaching CFP in Sierra Juárez, Baja Calif.)

Tragia (Euphorbiaceae: 100, tropical and subtropical, 1 in deserts of Calif., reaching CFP in Sierra Juárez, Baja Calif.)

Trichocoronis (Asteraceae-Eupatorieae: 2, S.W. North America, 1 widespread species in S. Calif., where it may not be native; R. M. King, personal communication)

Vauquelinia (Rosaceae: 8, S.W. U.S., Mexico; 1, N.W. Baja Calif.)

Verbena (Verbenaceae: 250, tropical and temperate America, 2-3 in Old World, 6 in Calif. and in CFP, 3 endemic)

Washingtonia (Arecaceae: 2, Calif., Ariz., Baja Calif.; 1 in deserts of Calif., and formerly in Palm Valley, 40 km E.S.E. of Tijuana, Baja Calif.)

Wislizenia (Capparaceae: 1, S.W. U.S. to Central Valley of Calif., adjacent Mexico)

Yucca (Agavaceae: 40, mainly in arid regions of North America; 4 in Calif., 1 common in CFP, a second reaching CFP in San Diego Co.)

Among these are approximately 241 annuals, about 40% of the total. There are other groups of plants, not listed in table 8, which from what might be considered Arcto-Tertiary or Madrean origins have radiated extensively into expanding semiarid regions and subsequently into the area of the California Floristic Province. Examples are *Astragalus, Cor-*

dylanthus, Penstemon, Phacelia, and some groups of *Allium* and *Galium,* to name just a few.

Malvaceae are particularly important in this group, since they seem to have radiated mainly from semiarid tropic-margin situations. Three genera in California—*Eremalche, Malvella* (including "*Sida hederacea* (Dougl.) Torr.";' Fryxell, 1974), and *Sphaeralcea*—are clearly desert elements that have come into California from the south and east. The last two of these are widespread elsewhere and not represented by endemics in California, whereas *Eremalche,* a genus of four annual species, has two endemics in and near the Central Valley of California and the other two in the adjacent deserts. One of the four species of *Malvella* is found in the Mediterranean region, whereas the other three are widespread in the warmer regions of the New World. *Hibiscus,* a genus of some 300 species of tropical to subtropical areas, has one widespread species in marshy places in the Central Valley and another in the deserts. Two other genera, *Abutilon* and *Horsfordia,* do not enter the California Floristic Province in California (*Abutilon* does in Baja California) but occur in the deserts of the State. The *Phrymosia-Malacothamnus-Iliamna* group (Bates, 1963, 1968; Bates and Blanchard, 1970; Fryxell, 1971), has apparently radiated from such origins to reach first Madrean and then Arcto-Tertiary vegetation. *Lavatera,* with four species on the islands off southern California and Baja California and all others in the Old World, is apparently a Madrean-Tethyan link. The ancestors of the fairly large genus *Sidalcea,* which is related to Eurasian genera such as *Lavatera* and *Malva* (D. M. Bates, personal communication), may have come through the north. *Sidalcea,* a New World genus, is mostly associated with derivatives of the Arcto-Tertiary Geoflora, although some species, including the only annuals in the genus, are in Madrean vegetation and endemic to California.

A similar pattern is shown by Euphorbiaceae. Of the 10 genera and 41 species in California, 9 genera and 19 species are in the California Floristic Province, with only 7 of the species endemic. The monotypic, near endemic, annual *Eremocarpus* is doubtless derived from *Croton* (Webster, 1967, p. 354), a genus that with *Acalypha, Chamaesyce, Euphorbia,* and *Stillingia,* is clearly marginal in California; *Tetracoccus* is endemic to the deserts of the southwestern United States and northwestern Mexico, just reaching the California Floristic Province. Euphorbiaceae exhibit no real evidence of connections with Eurasia via a northern or even a Madrean-Tethyan connection. On the other hand, Chenopodiaceae are listed here arbitrarily and many or all of the California genera may ultimately have come from the north.

Since wet and dry aspects of the vegetation have been in existence for a very long period of time, groups such as Polygonaceae, subfamily Eriogonoideae, which now seem to center primarily in semiarid and arid regions, have existed in such regions and also in those dominated by broad sclerophyll vegetation since early Neogene time. There seems little doubt that the center of evolution of many of the groups listed in table 8 has been in areas of vegetation drier than those now characteristic of the California Floristic Province, and that they have undergone considerable evolutionary radiation in this area since mid-Pliocene time, probably radiating into it chiefly during the Xerothermic periods. Certainly the low, warm deserts of North America were very restricted in area during the Pleistocene pluvial cycles, and have expanded greatly during the past 10,000 years. This trend toward increasing aridity has profoundly affected the flora of California as a whole (Axelrod, 1966b).

Cosmopolitan Elements

A few groups, mainly aquatics and other wetland plants, are so widespread that they do not fit easily into any of the groups reviewed above. These taxa comprise a total of 17 of the 154 native families in California, together with two additional genera, and include 29 genera and 92 species (table 9). There are two endemic species of *Callitriche,* one of *Elatine,* and one of *Isoetes.* These cosmopolitans comprise only about 2% of the flora of the State, but 11% of the families.

TABLE 9
Cosmopolitan Taxa in the Flora of California

Alismataceae	Marsileaceae
Callitrichaceae (2 endemic species)	Najadaceae
Ceratophyllaceae	Pontederiaceae
Elatinaceae (1 endemic species)	Potamogetonaceae
Haloragaceae	Ruppiaceae
Hydrocharitaceae	Typhaceae
Isoetaceae	Zannichelliaceae
Juncaginaceae	*Eclipta* (Asteraceae-Heliantheae)
Lemnaceae	*Zostera* (Zosteraceae)
Lentibulariaceae	

Relationships with South America

Many families of flowering plants ultimately seem to have come from South America to North America (Raven and Axelrod, 1974, p. 627-628). Examples from the flora of California include Amaryllidaceae-Allieae, Cactaceae, Fabaceae-Mimosoideae, Loasaceae, Nyctaginaceae, and Zygophyllaceae. In addition, there are many other families in which some elements appear to have come from South America, others from Eurasia; among them are Anacardiaceae, Asteraceae, Euphorbiaceae, Gentianaceae, Rhamnaceae, Solanaceae, and Rubiaceae in the flora of California. The entire order Chenopodiales (Centrospermae) may have originated in West Gondwanaland (= Africa + South America), in which case all of the constituent families are ultimately to be traced there (Raven and Axelrod, 1974). Unfortunately, the indications of such ancient events cannot be interpreted at present owing to the lack of an adequate fossil record. Some of the genera in California seem to have originated in South America and to have come to North America and to California secondarily (table 10). They include *Bowlesia, Cardionema, Paronychia,* and *Soliva,* which may have been introduced through human activities. The other genera enumerated in table 10 probably appeared in North America during the Quaternary also, even though human interference may not have played a role in their intercontinental transfers.

There are many species and species-pairs common to western North and South America (Raven, 1963). Most of them are self-compatible or autogamous herbs of open habitats that achieved their disjunct ranges after the establishment of mediterranean climates following the Late Pleistocene (Moore and Raven, 1970; Raven, 1972, 1973; Axelrod, 1973). The majority of these species or species-pairs have been derived from the north. The presence of relatively unspecialized species or groups of species of *Boisduvalia, Chorizanthe* (Reveal, 1977), and *Plagiobothrys* in South America probably reflects more than one time of arrival, rather than an origin there.

TABLE 10
Some Plants That Are Probably of South American Origin in the Flora of California

Acaena (Rosaceae: 100, South America, Australasia, 1 South Africa, to mountains of Mexico, 1 endemic to CFP; also Hawaii)

Bowlesia (Apiaceae, subfamily Hydrocotyloideae: 14, all South American, but with 2 probably naturalized species in North America, 1 of which grows elsewhere and in Calif.; Constance, 1963; Mathias and Constance, 1965)

Calandrinia (Portulacaceae: 150, mainly W. South America, relatively few in North America; reaches Australia. The whole *Talinum-Anacampseros-Ceraria-Lewisia-Lenzia* group, to which *Calandrinia* is related, or even the whole family, may ultimately have been derived from South America, but the 4 Calif. species of *Calandrinia* seem to be more recently derived, and 1 species is common to W. North America and Chile.)

Cardionema (Caryophyllaceae: 6, Pacific coast of North and South America, ours the only 1 in North America, described from Chile, Wash. to Mexico along coast; introduced in North America?)

Carpobrotus (Aizoaceae: 24, Australasia, South Africa, Chile, Calif., the 1 Calif. species common to Chile and Australia, *fide* S. T. Blake)

Dichondra (Convolvulaceae: 15, wide distribution, 2 in Calif., endemic to CFP, perhaps derived from South America; Tharp and Johnston, 1961)

Dissanthelium (Poaceae: 17, 16 in the Andes of South America with 1 perennial species N. to high mountains in Central Mexico; the single annual species in Calif., probably extinct, once grew on Santa Catalina, San Clemente, and Guadalupe Isds. [Swallen and Tovar, 1965]; and its relationships to the rest of the genus should be investigated in detail)

Gaultheria (Ericaceae: 200, Himalayas and S. India to Australasia and South America, Mexico, 2 in E. North America, 3 in CFP)

Larrea (Zygophyllaceae: 5, Cent. W. and S. South America, 1 in deserts of North America to S. San Joaquin Valley and cismontane Riverside Co. May have reached North America only in Late Pleistocene time.)

Lepechinia (Menthaceae: 40, South American and Mexican, 4 in CFP, also 1 in Hawaii; Epling, 1948)

Limonium (Plumbaginaceae: 300, 1 in Calif., monomorphic and self-compatible, closely related to a heteromorphic, self-incompatible Chilean species; Baker, 1953; Raven, 1963)

Nicotiana (Solanaceae: 70, about 40 in South America, about 20 in Australasia; 1 in South West Africa, 9 in North America, 4 in Calif., 1 endemic to and 1 reaching CFP; Goodspeed, 1954)

Odontostomum (Liliaceae: 1, endemic to mountains bordering Sacramento Valley. According to Hutchinson, 1973, the only Northern Hemisphere representative of Tecophilaeaceae, with 3 additional genera in Chile and 2 in Africa.)

Paronychia (Caryophyllaceae: 50, at least 21 in North America; Core, 1941. Other important centers of differentiation are in western Eurasia, where there are 11 species, North Africa, and S. South America. The 1 species in Calif. has a very restricted range along the coast in San Francisco, Marin, and Sonoma counties, does not appear closely related to any other North American species, and might even have been introduced from Chile. Its affinities should be investigated further.)

Pityrogramma (Pteridaceae: 40, tropical America and Asia, 1 polymorphic assemblage in W. temperate North America; Alt and Grant, 1960; Smith et al., 1971)

Soliva (Asteraceae, tribe Anthemideae: 8, all in South America unless *S. daucifolia* Nutt. is distinct from *S. sessilis* Ruiz & Pav., in which case it may be a Calif. endemic and both are in Calif.; differences probably maintained by autogamy. *Soliva pterosperma* (Juss.) Less. is common to CFP and Argentina. Probably all are introduced in North America; Crampton, 1954; Raven, 1963, p. 175.)

Relationships with the Mediterranean Region

There are a limited number of similarities between the floras of the Mediterranean region proper and California (Raven, 1973; table 5). The species of some of these genera seem to have had their origin from taxa in a common northern flora; for example, *Acer; Alnus; Betula; Clematis; Crataegus; Fraxinus; Juniperus* subg. *Juniperus* and perhaps also subg. *Sabina; Lonicera; Populus; Prunus* subg. *Cerasus, Padus,* and *Prunus; Rhamnus; Rosa; Rubus; Smilax; Staphylea; Viburnum;* and *Vitis.* Other genera (table 5) may have migrated more directly between semiarid or subhumid regions in Europe and North America via a more southerly route, in the broad ecotone between the Arcto-Tertiary province and the dry climates to the south; such links are termed Madrean-Tethyan (Axelrod, 1975). In a few genera—*Galium; Juniperus* subg. *Sabina, Populus;* and *Pinus* and *Quercus* are possible examples—there may be both northern and Madrean-Tethyan links, and some, like *Celtis,* appear to be intermediate.

There were more numerous links between California and the Mediterranean region in the Miocene and Pliocene, among them the following genera: *Clethra, Ilex, Ocotea, Persea, Pistacia, Sabal, Sageretia, Sapindus, Zanthoxylum,* and *Zizyphus* (Axelrod, 1975), but these became extinct in California as summer rains were reduced and the woody flora was reduced in diversity. All of these genera survive in Mexico and some also persist in the southern United States where summer rainfall is adequate.

A few species-pairs are in California and in the Mediterranean region (Raven, 1973, p. 217). In view of the many weeds that have originated in the Mediterranean Basin, where agriculture is of antiquity (Naveh, 1967), it would not be surprising if some of the supposed links may have had their origin in human introduction. Among the possible examples are *Plantago ovata* Forssk. (Stebbins and Day, 1967; Bassett and Baum, 1969) and *Oligomeris linifolia* (Vahl) Macbr., the only species of Resedaceae in the New World.

The other groups should be investigated in detail. The patterns are unclear in some other genera that seem to have comparable disjunctions in range, including *Aphanes; Caucalis; Valerianella-Plectritis;* and perhaps *Lupinus* and *Erysimum* (Meusel, 1969); the trans-Atlantic relationships in these groups ought to be clarified. The far greater number of disjunct species between areas of mediterranean climate in California and Chile than between California and the Mediterranean region itself may be related to the regular paths of bird migration between the two former areas (Raven, 1973).

In several genera, among them *Antirrhinum, Astragalus, Atriplex, Bromus, Juncus* (Hermann, 1948), *Poa, Polygonum, Senecio, Silene,* and *Trifolium,* annual species seem to have been independently derived from perennial species in the Old World and in the New. This is a significant relationship that should be investigated further in a comparative manner.

Summary

Earlier workers who dealt with the flora of California (e.g., Abrams, 1925, 1926; Jepson, 1925; Campbell and Wiggins, 1947), have stressed its unique character. This is not surprising since 52 genera—6.5% of the total—and about 48% of the species are endemic in the California Floristic Province. Taking a broader view, however, and looking for ultimate sources, it is clear that the flora of California can be thought of as comprising two elements: a northern, or Arcto-Tertiary, and a southern, or Madro-Tertiary, one (Axelrod, 1958), a relationship clearly perceived by Asa Gray nearly a century ago (Gray, 1884). Within the area of the broad ecotone between these fundamentally different vegetation types, a mediterranean climate developed following the Pliocene, and provided a major stimulus for the proliferation of species and probably some genera, for example Asteraceae and Scrophulariaceae (Stebbins and Major, 1965; Lewis, 1972). During arid phases of the Quaternary, plants from the drier parts of the Madrean region entered the California Floristic Province, especially in the drier valleys of cismontane southern California, the San Joaquin Valley, and the inner South Coast Ranges, and have evolved endemic species there.

More than half of the genera and species in the California Floristic Province have Arcto-Tertiary affinities. Another quarter of the genera and a third of the species are primarily associated with plant formations derived from the Madro-Tertiary Geoflora. Twenty percent of the genera and 15 percent of the species have migrated into the region of the California Floristic Province from the deserts in fairly recent time, perhaps mostly during and since the Xerothermic period about 8,000-4,000 years B.P. The remaining 5% of the genera, with a small number of species, come from miscellaneous sources.

TRANSMONTANE CALIFORNIA: THE DESERTS

Thus far, our discussion has been focused on the California Floristic Province, an area of great importance and interest. About 744 of the 878 genera (85%) and 4119 of the 5046 species (82%) found in the State occur in this region, which comprises about 75% of the total area of California. We shall now focus on the approximately 102 genera and 935 species that occur only east of the mountains and on the deserts, but within the boundaries of California. Here are our only representatives of the families Burseraceae, Elaegnaceae, Koeberliniaceae, Krameriaceae, Rafflesiaceae, and Simaroubaceae—six of the 154 total native families in California. Two additional families that occur in the deserts of California but enter the California Floristic Province in Baja California are Arecaceae and Fouquieriaceae. The 121 genera found within the borders of the State but not in the California Floristic Province are listed in table 11.

History

The fossil record shows that the regional deserts now at the margin of the California Floristic Province are recent in origin. The Great Basin Desert (Shreve, 1942) occupies an area that was covered by mixed conifer forest and broadleaved sclerophyll vegetation into the Late Miocene (12-13 m.y.), and by juniper-oak-pinyon woodland during the Pliocene. The taxa that contribute to the present desert were either represented then by taxa that still occur at the drier margins of forest and woodland, or were derived directly from them. Clearly, they were largely preadapted to the drier conditions that spread widely over the lowlands as the Sierra-Cascade barrier was elevated in the late Cenozoic and a drier climate spread in its lee following the pluvial periods. Desert environment alternated with non-desert conditions in response to fluctuating glacial-nonglacial climates. Late Pleistocene pollen records indicate that the northwestern part of the province was covered with scattered *Pinus ponderosa*, with patches of *Artemisia* and its associates in the region. Plant remains in wood-rat middens, which provide more definitive evidence of the actual vegetation in the immediate area, demonstrate that pinyon-juniper woodland reached down to the floor of the present southern Great Basin Desert in the pluvial stages when precipitation at a minimum was 250-260 mm greater than at present.

Some genera and pairs of species provide links between the Great Basin and the steppes of Asia, among them *Artemisia, Aster, Atriplex, Ceratoides* (*Eurotia;* Howell, 1971), *Chamaerhodos, Crepis, Lactuca, Stipa,* and *Suaeda* (Babcock and Stebbins, 1938; Axelrod, 1950, p. 264-265; Yurtsev, 1972). The ancestors of these plants may well have spread between the Old World and the New on more than one occasion. Judged from the influx of grazing

TABLE 11
Genera Found in California but Not in California Floristic Province

Acamptopappus	*Leucelene*
Acleisanthes	*Lippia* s. str.
Ammoselinum	*Lygodesmia*
Amphipappus	*Lyrocarpa*
Amsonia	*Malperia*
Androstephium	*Matelea*
Arctomecon	*Maurandya*
Atrichoseris	*Mohavea*
Ayenia (Cristóbal, 1960)	*Monarda*
Blepharidachne	*Monoptilon*
Brandegea	*Mortonia*
Bursera	*Neolloydia*
Calliandra	*Nicolletia*
Canbya	*Olneya*
Canotia	*Oryctes*
Carnegiea	*Oxystylis*
Cassia	*Palafoxia*
Castela (*Holacantha*)	*Peraphyllum*
Cercidium	*Petalostemum*
Chaetadelpha	*Petradoria*
Chamaesaracha	*Petrophytum*
Cleomella	*Peucephyllum*
Coldenia	*Physaria*
Coleogyne	*Pilostyles*
Colubrina	*Pleurocoronis* (*Hofmeisteria* in part)
Coryphantha	*Polanisia*
Dedeckera (Reveal and Howell, 1976)	*Poliomintha*
Dicoria	*Polyctenium*
Dimeresia	*Prenanthella*
Ditaxis	*Psathyrotes*
Eatonella s. str.	*Psilostrophe*
Echinocactus	*Purpusia*
Enceliopsis	*Ratibida* (Ediger and Santamaria, 1971)
Enneapogon	*Salazaria*
Euphorbia subg. *Poinsettia*	*Sarcobatus*
Fallugia	*Schizachyrium*
Forsellesia	*Sclerocactus*
Gilmania	*Scophulophila*
Glyptopleura	*Selinocarpus*
Grayia	*Shepherdia*
Gymnosteris	*Stanleya*
Halimolobos	*Swallenia*
Hecastocleis	*Teucrium*
Hesperocallis	*Tidestromia*
Hilaria	*Townsendia*
Horsfordia	*Trianthema*
Hymenoxys	*Tricardia*
Kallstroemia	*Trichoptilium*
Koeberlinia	*Tripterocalyx* (Galloway, 1976)
Krameria	*Tridens*
Laphamia	*Xylorhiza*

mammals in the Middle Pliocene, some taxa may have been carried across at that time. This need not have been continuous animal transport, but in steps that provided a spotwise distribution, much as the *Juniperus communis-Artemisia frigida* Willd. association links these desert-border areas today. Closer dating of the times of migration across the region must await the discovery of fossil floras in the Beringian area that clearly indicate steppe and steppe-border conditions.

The fossil record shows that during the Miocene and Pliocene the area of the present Mohave and Sonoran deserts in California was blanketed with a rich live oak-laurel-palm woodland filled with numerous sclerophyllous shrubs. The occurrence of a very similar flora in each province indicates that they formed a phytogeographic unit at that time, and that the present Mohave and Sonoran desert floras have been differentiated more recently. This is consistent with geologic evidence which shows that the Mohave province was uplifted during the Late Pliocene-Pleistocene, bringing to it a colder climate which eliminated many thorn scrub and other taxa from that region, confining them to the south. Among them were species of *Bursera, Cercidium, Colubrina, Condalia, Lycium,* and *Washingtonia* (Axelrod, 1950, p. 271). Significantly, a few of these, including *Acacia, Chilopsis, Dalea,* and *Prosopis,* still have relict outposts in the Mohave province, notably in the low-lying, warmer Ludlow-Amboy basins that reach northward into the Mohave region (Axelrod, 1950, p. 271). As colder climates spread across the new uplifted Mohave province it was stocked by taxa from the Great Basin area to the north, including species of *Artemisia, Purshia,* and many others. During the moister phases, pinyon-juniper woodland, now greatly impoverished, descended into the lowlands from the bordering ranges, as revealed by their remains in wood-rat middens in the region. The absence or poor representation of many life-forms and low percentage of endemism in the Mohave Desert (no genera, 3.3% of the species) reflects its recent and rapid uplift and the resulting development of a colder climate there.

The Sonoran Desert, which extends into the southeast corner of the State, is a subtropical desert, filled with numerous small trees, cacti, shrubs and half-shrubs (Shreve, 1942). They include a great variety of life forms, and numerous isolates in diverse families that obviously have had a long history. This no doubt took place on the drier margins of savanna, thorn forest and woodland vegetation, where most of these taxa have allies today. Evidently they were gradually derived through time, and have been selected for the increasingly drier climates which developed over the area. Among the older isolates that may well be at least of Paleogene age are the genera *Crossosoma, Fouquieria,* and *Koeberlinia.* A number of families that are represented in Late Cretaceous and Paleocene floras have representatives in the desert province today. These include unique, isolated taxa that occur in thorn forest, short tree forest, and subtropical forest. We may infer that some of them, for instance *Dasylirion* and *Nolina* (Liliaceae), *Viscainoa* (Zygophyllaceae), *Castela* (including *Holacantha*; Simaroubaceae), *Simmondsia* (Buxaceae), *Pachycormus* (Anacardiaceae), *Atamisquea* and *Forchammeria* (Capparaceae), *Olneya* (Fabaceae), and *Choisya* (Rutaceae), were present in drier sites over the region by the early Tertiary. A number of genera that are recorded in Late Cretaceous to Eocene floras, where they are in tropical savanna to thorn forest vegetation, may well have been present over the area of southeastern California by the Eocene, including *Ficus* and *Morus* (Moraceae); *Acacia, Cassia, Caesalpinia,* and *Pithecellobium* (Fabaceae); *Colubrina* and *Zizyphus* (Rhamnaceae); *Cardiospermum* and *Sapindus* (Sapindaceae); *Bumelia* and *Sideroxylon* (Sapotaceae), *Bombax* (Bombacaceae), and others. On this basis, the rather distinct species of these genera may well have appeared by the Late Eocene-Oligocene. During the middle Tertiary it is apparent that distinct species of numerous genera were originating, in *Abutilon, Agave, Atriplex, Bursera, Dalea, Euphorbia, Ipomoea, Jatropha, Krameria, Lycium* and *Yucca.* It was chiefly during the fluctuating glacial-interglacial ages that numerous species of many genera, especially herbaceous ones, originated in the region. In addition, many mesic taxa that range far out of the desert province today may well have entered it during the last pluvial, and were "stranded" in locally favorable habitats as the climate became progressively more arid. Thus, the richness of the desert flora owes chiefly not to the antiquity of the desert on a regional scale, but rather to the accumulation of numerous taxa during the Tertiary and Quaternary, taxa that were preadapted to increasing drought over the region.

Analysis

As pointed out by Stebbins and Major (1965), and in fig. 7, there are four distinct floristic regions in eastern and southeastern California outside of the California Floristic Province, with different affinities outside of the State. Each of these units will be discussed in turn.

Great Basin.—In the northeast corner of the State, this region comprises all of Modoc County and portions of Lassen, Shasta, and Siskiyou counties, and also includes northernmost Inyo County above the Owens Valley and Mono County east of the Sierra Nevada. Although a number of taxa of the California flora extend northeastward beyond the main Sierra-Cascade axis to the Warner Mountains and other portions of the Modoc Plateau (e.g., Griffin, 1966), the flora of that region, which comprises some 22,600 km^2, is essentially that of the Great Basin. Unfortunately, only a portion of its northern, larger part is included in the Intermountain Flora (Cronquist, et al., 1972), and the Warner Mountains, which certainly belong to the Intermountain region, were excluded from that work. The range is one of the more interesting areas phytogeographically within the State, and would make an excellent subject for a local floristic study.

Within the northern part of the Great Basin region in California are some 127 species that do not enter the California Floristic Province, including 67 that are found nowhere else in the State. In the southern part of the same region, there are about 147 species that do not occur in the California Floristic Province, together with 26 that are found nowhere else in the State, including the genus *Chaetadelpha. Lomatium ravenii* Math. & Const. and *Mimulus pygmaeus* Grant are the only endemics known to us to be restricted to the northern part of the Great Basin region within the state of California. There are at least seven endemics within the southern part, however, together with at least seven others endemic in California and five others that range a short distance into Nevada. Because there is topographic continuity between the White Mountains and the Inyo Mountains, we have used the boundaries of the Inyo and Great Basin regions elastically in making up the preceding statistics.

The White Mountains, which lie at the southwestern corner of the Great Basin region (Stebbins and Major, 1965; Cronquist et al., 1972), have been the subject of a valuable recent flora by Lloyd and Mitchell (1973). Following a detailed analysis of the 763 native species and varieties, the authors conclude that the White Mountains are a Great Basin desert range enriched during the Pleistocene with many boreal taxa via the Sierra Nevada. They are especially abundant in alpine and subalpine areas. Such families as Fagaceae and

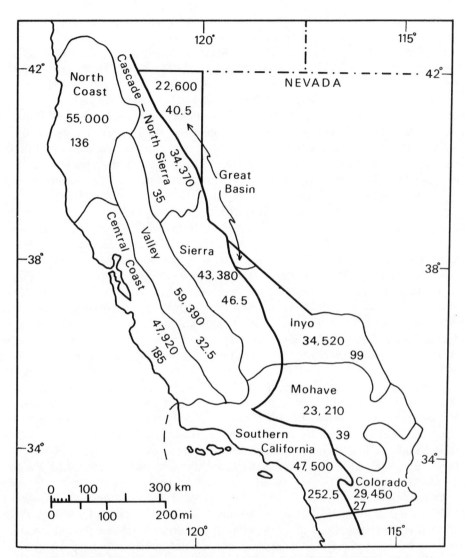

Fig. 7. Floristic regions of California. The larger figure in each region represents its area in km². The smaller figure indicates the number of endemics in 70 large- and intermediate-sized genera. The area west of the heavy line is the California Floristic Province as defined in the text. From Stebbins and Major (1965).

Ericaceae, though common in the Sierra Nevada, are absent in the White Mountains. In addition, many Sonoran-Mohavean (= Madrean) taxa have come from the south, especially in latest Pleistocene and Recent time.

Inyo Region.—The Inyo region (fig. 7) is the richest and most interesting in transmontane California. About 455 species that occur in this region do not reach the California Floristic Province, including at least 200 that are not found elsewhere in the State. There are at least 44 endemic species within the Inyo region, with 10 other species endemic in California but not found within the California Floristic Province, and more than 25 additional species endemic to this area that range a short distance into Nevada. The following genera are endemic to the Inyo region, some ranging out of the state of California: *Arctomecon* (Papaveraceae: 2), *Dedeckera* (Polygonaceae: 1; Reveal and Howell, 1976), *Gilmania* (Polygonaceae: 1), *Hecastocleis* (Asteraceae, tribe Mutisieae: 1), *Oxystylis* (Capparaceae: 1), *Scopulophila* (Caryophyllaceae: 1), and *Swallenia* (Poaceae: 1; Soderstrom and Decker, 1963). In addition, there are many peculiar and isolated species, such as *Eriogonum intrafractum* Cov. & Mort. (cf. Reveal, 1969), *Boerhavia annulata* Cov., and *Tetracoccus ilicifolius* Cov. & Gilman. Some of the endemics of the Death Valley region may have evolved rapidly, as discussed by Iltis (1957) for *Oxystylis*, whereas others have probably developed over long periods of time on the limestone cliffs or migrated into the region from the south following the last pluvial period. Endemism on limestone in the deep canyons flanking Death Valley is a notable feature of the region (see pp. 68-69).

As pointed out by Stebbins and Major (1965, p. 9), the Inyo region is distinctive mainly because of the flora of its high mountain ranges. These include the White, Panamint, Argus, Kingston, Clark, New York, and Providence Mountains. Many of the unusual plants of this area, like those discussed above, are endemic. A number of others represent western outposts of the Rocky Mountain flora, including (among the genera that occur in California only in the Inyo region) *Buddleja*,[1] *Chamaesaracha* (excluding *Leucophysalis*), *Cowania, Fallugia, Monarda, Mortonia,* and *Physaria.* Other Rocky Mountain and Great Basin genera in this region that have wider distributions in California include *Chamaebatiaria, Jamesia, Pericome,* and *Purshia,* as well as such species as *Pinus flexilis* James and *P. longaeva* D. K. Bailey. The presence of many such species in the Providence, New York, Clark, and Kingston Mountains especially (Henrickson and Prigge, 1975) is indicative of greater amounts of summer precipitation than at lower elevations, and perhaps also of enhanced effective moisture where the precipitation runs off smooth-faced cliffs into crevices, as noted by Danin (1972) and by Danin, Orshan, and Zohary (1975) for the climatically similar mountains of the eastern Mediterranean region. Among the other genera that occur in California only in the Inyo region are *Enceliopsis, Enneapogon, Gaura, Halimolobos, *Hedeoma, Laphamia, Maurandya, Petradoria, Petrophytum, Purpusia, *Sanvitalia, Selinocarpus,* and *Tragia.*

Mohave Desert.—In the Mohave Desert, as discussed earlier, there are about 757 species, of which about 22 are endemic; 6 others are not found in the California Floristic Province but occur in other regions east of the mountains, and at least 7 others are essentially endemic to the Mohave Desert but range into Nevada or northwestern Arizona. About 51 of the species in the Mohave Desert are not found in any other part of the State. Approximately half of the species in the Mohave Desert extend into the California Floristic Province, reaching to the San Joaquin Valley, the mountains of southern California, or the inner South Coast Ranges. The lines between the Mohave region and the Inyo and Sonoran Desert regions are gradual and complex transitions, so it has been difficult to assign some species to these zones with precision. A flora and detailed analysis of the Mohave Desert as a whole, including those portions which lie in Nevada, Utah and Arizona, would be of great value. A most useful recent flora, which includes the transition zone between the Great Basin and Mohave Desert in south-central Nevada, is that of Beatley (1976).

---

[1] The genera marked with an asterisk (*) also occur in northwestern Baja California within the limits of the California Floristic Province.

Sonoran Desert. —The "Colorado Desert" is the northwestern arm of the Sonoran Desert which lies in California (Shreve and Wiggins, 1964). It has about 306 species not found in the California Floristic Province, including about 100 that do not occur elsewhere in the State. There are at least eight endemic species, together with two others endemic to the Colorado and Mohave Deserts. The lack of summer rain over much of the Mohave Desert, and particularly the lower winter temperature, restricts a number of desert plants in the California portion of their range to the Colorado Desert, among them the following genera: *Acleisanthes, Ammobroma, Ammoselinum, Bernardia, Bursera, Calliandra, Carnegiea, Cercidium, Colubrina, Condalia, Fouquieria, Hyptis, Koeberlinia, Lyrocarpa, Malperia, Olneya, Pilostyles, Schizachyrium, Sesbania, Teucrium,* and *Washingtonia.* A number of other genera, including *Achyronychia, Cassia, Chilopsis, Dalea, Ditaxis, Hoffmanseggia, Krameria,* and *Xylorhiza* in California, are most characteristic of the Sonoran Desert but range northward to the Mohave Desert or in a few instances to the desert of the Inyo region as well. Of the genera just mentioned, however, *Achyronychia, Bernardia, Condalia, Dalea, Fouquieria, Hoffmanseggia, Hyptis,* and *Washingtonia* occur within the borders of the California Floristic Province in Baja California. A very useful atlas of plant distributions for the Sonoran Desert is that of Hastings, Turner, and Warren (1972).

Summary

Approximately 6 families, 102 genera, and 938 species of vascular plants occur within the boundaries of the state of California but are not in the California Floristic Province. Despite the fact that all areas east of the mountains in California intergrade with and form integral parts of larger floristic regions that lie mainly outside of the State, there are three endemic genera—*Dedeckera, Gilmania,* and *Swallenia*—and about 85 endemic species in this region. In addition, at least four other genera—*Arctomecon, Hecastocleis, Oxystylis,* and *Scopulophila*—and a minimum of 25 other species endemic in the Death Valley region or the Mohave Desert extend only a short distance into southern Nevada or northwestern Arizona.

The most important center of endemism east of the mountains in California is the Inyo region. A detailed floristic analysis of this region would be of high interest and utility. Two other areas deserving more detailed study are the Warner Mountains and the Mohave Desert. In both regions the composition of their floras reflects a mixture of "Californian" and "extra-Californian" species that has not yet been analyzed in detail.

SIX FAMILIES PROMINENT IN THE CALIFORNIA FLORA

As examples of the patterns we have been discussing, the composition and probable evolutionary history of a few of the families that contribute in a very significant way to the flora of California are traced. These families all contain a high proportion of endemic species. Each is best developed in the California Floristic Province.

Amaryllidaceae, tribe Allieae

The tribe Allieae of Amaryllidaceae (Hutchinson, 1973, p. 793) consists of three groups (Moore, 1953; Traub, 1963). Eleven genera, comprising the subtribe Alliinae, have tunicated bulbs. *Allium* itself, with some 450 species, is far better represented in the Old World than in the New. The other 10 genera in this group are *Milula* (1 species, Himalayan), *Nec-*

taroscordum (4 species, W. Mediterranean to Iran and the Caucasus); *Caloscordum* (2 species, Asia); *Ipheion* (25 species, Mexico to Chile); *Leucocoryne* (*Latace*, 15 species), *Garaventia* (*Steinmannia*: 1 species), *Stemmatium* (1 species), and *Tristagma* (30 species), all Chilean, with *Tristagma* reaching Patagonia; *Nothoscordum* (18, Argentina to Chile, with other species in south-central United States); and *Tulbaghia* (24 species, Africa). A second group of seven genera, the subtribe Brodiaeinae, consists of species with corms that have coarsely fibrous-reticulate tunics, and includes five genera—*Bloomeria, Brodiaea, Dichelostemma, Muilla,* and *Triteleia*—that center in the California Floristic Province, with 39 of the 42 total species found there and 32 of them endemic. Two other genera, *Androstephium* and *Triteleiopsis,* with a total of three species, are endemic in the adjacent deserts. The third subtribe, Millinae, consists of seven genera. They are all Mexican and Central American, and comprise *Milla* (6 species), *Dandya* (3, Lenz, 1971), *Diphalangium* (1), *Petronymphe* (1), and *Bessera* (2; Moore, 1953). A South American (or South American + African) origin appears likely for the tribe Allieae (Raven and Axelrod, 1974), and the degree of radiation of the subtribe Brodiaeinae in the California Floristic Province suggests a relatively great antiquity for that group in Madrean vegetation. Its relationships with *Allium,* however, are doubtful and the similarity of these groups might be attributable to convergent evolution toward the same habit, as suggested by the late L. Mann (R. Y. Berg, personal communication).

Boraginaceae

This very large family of some 115 genera and 2400 species is world-wide but best developed in the Old World, especially in the Mediterranean region. In California, there are 13 genera and 145 species, 45 endemic; in the California Floristic Province, 13 genera and 111 species, 68 endemic. The greatest proliferation of species and of endemism in California occurs in the closely related tribes Cynoglosseae and Eritrichieae, both basically of north temperate orientation. Characteristic of the former are the scarcely distinct white-flowered annual genera *Harpagonella* (1; S. Calif. to central Baja Calif. and Ariz.) and *Pectocarya* (11; disjunct between W. North America and Chile, with 4 endemic to South America, 5 to North America, and 2 common; Raven, 1963, p. 174). The great majority of genera and species of the tribe Cynoglosseae are endemic to Eurasia, and there seems no doubt that *Harpagonella* and *Pectocarya* have been derived from Arcto-Tertiary ancestors that entered the drier Madrean vegetation of western North America, reaching South America secondarily. *Pectocarya* is best represented in the deserts and most species probably reached the area of the California Floristic Province secondarily, possibly during the Xerothermic.

Eritrichieae center around a complex of north temperate genera such as *Lappula, Eritrichium,* and *Hackelia,* together with a number of other small Asian genera. In western North America other genera are *Cryptantha,* with nearly 100 species, 61 in California; *Plagiobothrys,* with about 100 species, 39 in California, and *Amsinckia,* with about 18 species, 11 in California. All three genera are likewise found in temperate South America (Raven, 1963, p. 174); all perennial species of *Cryptantha* and all heterostylous species of *Amsinckia* are North American, but seven of the nine perennial species of *Plagiobothrys* are South American. All three genera doubtless had a West North American origin.

Of the 61 species of *Cryptantha* in California, 41 are in the California Floristic Province and 22 of them are endemic. *Plagiobothrys* has 34 species in the California Floristic

Province, including 22 endemics, and *Amsinckia* all 11 species, including five endemics. *Plagiobothrys* sect. *Allocarya* consists of about 30 annual species endemic to western North America, about six endemic to South America, and at least three in Australia. Of the perennial species, however, seven are endemic to South America, one to Mexico, and one to western North America—a pattern similar to that in *Chorizanthe* (Raven, 1963, p. 174). Of the approximately 30 North American species, 22 occur in the California Floristic Province and 17 of them are endemic. All three genera are closely associated with Madrean vegetation, and both *Amsinckia* (judged from the occurrence of primitive species) and *Plagiobothrys* sect. *Allocarya* seem to be centered in the California Floristic Province. These three genera contribute 47 endemics to the California Floristic Province, 44 of them annuals. Although Boraginaceae have only about 6% of their species in California, they contribute significantly to the endemism in this region.

Hydrophyllaceae

Fourteen of the 18 genera of Hydrophyllaceae and 130 of the approximately 250 species occur in California, including 39 endemic species and two endemic genera, both monotypic—*Draperia* and *Lemmonia*. In the California Floristic Province there are 13 genera and 99 species, including 65 endemics. The genera not found in California are the widespread tropical and subtropical *Hydrolea* (20 species); the South African *Codon* (2); and the tropical American woody *Wigandia* (6). In addition to the two endemic genera just mentioned, *Pholistoma* (3), *Eucrypta* (2), *Emmenanthe* (2), *Turricula* (1), and *Eriodictyon* (8) are restricted to the California Floristic Province and to some extent the deserts of the western United States and adjacent Mexico. *Tricardia* (1) occurs only on these deserts, whereas *Romanzoffia* (4), *Hesperochiron* (2), and *Hydrophyllum* (8) occur mainly in temperate forests, especially in western North America. *Nemophila* (13) has a similar distribution but also has four annual endemics in the California Floristic Province. *Nama* (45) is American (one species in Hawaii) and occurs mainly in semiarid regions. Two closely related perennial species, *N. lobbii* A. Gray and *N. rothrockii* A. Gray, that are endemic to the mountains of California and western Nevada, may not be closely allied to the others and probably constitute a distinct genus.

The largest genus of the family, *Phacelia*, with some 200 species of North and South America, has 89 species in California and 63 in the California Floristic Province, including 39 endemics. Some of the more generalized species of *Phacelia* (e.g., *P. dalesiana* J.T. Howell; Constance, 1952) occur in association with temperate forests, but many groups are now mainly in the deserts, and some of these have entered the California Floristic Province secondarily, probably in the Xerothermic. The western United States obviously constitutes an important area of evolution for the family, and many of its genera have been entirely or primarily associated with Madrean vegetation since their origin.

Onagraceae

Of the 17 genera and 650 species of Onagraceae, 10 genera and 135 species, with 43 endemics, are in California. Most of the least specialized species of *Epilobium* (including *Zauschneria*) occur together with *Boisduvalia* in the California Floristic Province, suggesting an origin for the tribe Epilobieae within the more mesic part of Madrean vegetation in western North America (Raven, 1976a). Of the 11 genera of the tribe Onagreae, with about 275 species, 6, with 104 species, are in California, including the endemic, monotypic *Heterogaura.* The annual genus *Clarkia* has 41 species, 39 found in California, and 36 endemic to the California Floristic Province; *Camissonia* sect. *Holostigma* (Raven, 1969) has 10 of its 12 species endemic to the California Floristic Province, one ranging beyond it and one reappearing in central Arizona. The main evolutionary radiation of the tribe Onagreae has taken place chiefly in Madrean vegetation in western North America.

Polemoniaceae

This family consists of 18 genera and about 317 species (Grant, 1959). Five of the less specialized genera, comprising three tribes and a total of 38 species, are Latin American, and the family may have originated in South America. The remaining 13 genera and approximately 278 species are temperate North American, centered in western North America. Included are two tribes, Polemonieae and Gilieae.

Polemonieae include six genera: *Polemonium* (23 species, 7 in California), *Allophyllum* (5-5), *Collomia* (14-8), *Gymnosteris* (2-1), *Phlox* (61-13), and *Microsteris* (1-1). One species of *Polemonium*, 10 of *Collomia*, four of *Phlox*, and all of *Allophyllum* and *Microsteris* are annuals. *Polemonium*, which is the least specialized genus, approaches in its characteristics the more primitive tropical genus *Cobaea* (Grant, 1959). The early history of Polemonieae seems clearly to have taken place in the temperate forests of North America, which their ancestors reached from the tropical antecedents. As aridity spread and new plant communities appeared in western North America, some members of the tribe radiated into them. Thus, California has been the major center of differentiation for the specialized, annual genus *Allophyllum*, and a secondary center for *Collomia.* In contrast, Gilieae, comprising *Gilia* (57-39), *Ipomopsis* (24-5), *Eriastrum* (14-14), *Langloisia* (5-4), *Navarretia* (30-29), *Leptodactylon* (6-3), and *Linanthus* (37-34) present a pattern similar to that to be discussed below for Polygonaceae, tribe Eriogoneae, with the least specialized genera such as *Ipomopsis* and *Leptodactylon* in less xeric Madrean vegetation, and approaching *Loeselia* of semiarid Madrean vegetation in Mexico (Grant, 1959). *Leptodactylon*, two species of *Linanthus*, one of *Eriastrum*, all but six species of *Ipomopsis*, and about a dozen species of *Gilia* are perennials, for a total of 39 of the 172 species of Gilieae. In California are 126 of the 172 species of Gilieae, of which all but 9 are annual.

There are 104 species of the tribe Gilieae in the California Floristic Province, 96 of them annuals, and 65 endemic. Particularly noteworthy is *Navarretia*, with all 29 of the North American species in the California Floristic Province; one additional species is South American (Crampton, 1954; Raven, 1963). Of these 29 species, 24 are endemic, the others ranging farther north. The genus appears to be a prime candidate for biosystematic attention. Most species of *Linanthus*, 32 out of 37, are also in the California Floristic Province, but a number of species occur on the deserts, as in the other genera of the tribe Gilieae. It appears that the tribe Gilieae evolved in response to increasing aridity in western North America, perhaps in the Miocene, but it may have had an older association with the broad sclerophyll Madrean vegetation than does the Polygonaceae, tribe Eriogoneae. For it, too, diversification in such vegetation has undoubtedly been spurred by increasing aridity following Early Pliocene time. Over half of the species of Polemoniaceae (163 of 317) are found in California, and more than 23% (74 species) of all the species in the family are endemic to the California Floristic Province.

Polygonaceae

This family consists of about 40 genera and 900 species. Of the three subfamilies (Dammer, 1889; Roberty and Vautier, 1964; Reveal and Howell, 1976), the presumably least specialized and primitively woody subfamily Calligonoideae, is centered in the Southern Hemisphere. In the well-known, primarily north temperate Polygonoideae, representatives of two tribes, Polygoneae and Rumiceae, occur as natives in California. Although *Polygonum* sens. lat. (300 species) is worldwide in distribution, one distinctive group (sect. *Duravia*), is centered in western North America. This section should be defined by pollen morphology, among other characters, to include *P. paronychia* Cham. & Schlecht. and *P. shastense* Brew. & Gray, as well as species no. 8-18 in Munz and Keck, 1959 (Hedberg, 1946; Mertens and Raven, 1965). In California, there are about 25 native species of *Polygonum*. The second tribe, Rumiceae, includes two native genera in California, *Rumex* (with 14 native species) and *Oxyria* (with 1) which is circumboreal.

The third subfamily, Eriogonoideae (Reveal and Howell, 1976; Reveal, 1977), consists of two tribes. Pterostegieae include two small genera: the shrubby or scandent ditypic *Harfordia* of Baja California, reaching the southern limits of the California Floristic Province, and the more widespread monotypic annual *Pterostegia* of the western United States and northwestern Baja California. Eriogoneae consists of 12 genera, which Reveal (1977) considers to represent three distinct clusters. Closely related to and probably derived from *Eriogonum* (247 species), the most primitive genus of the tribe, are *Oxytheca* (7 species, western U.S.; 1, Chile and Argentina); *Dedeckera* and *Gilmania*, monotypic genera of the Death Valley region; *Stenogonum*, two species of the Colorado Plateau and adjacent areas; *Goodmania* (Reveal and Ertter, 1977) and *Hollisteria*, monotypic genera endemic to central and southern California; and *Nemacaulis*, another monotypic genus of the southwestern United States and northwestern Mexico. In the second cluster, the basal genus appears to be *Chorizanthe*, comprising about 50 species of western North America (where they center in California to some extent) and Chile and Argentina. About 10 species are South American, and all but one of them are perennial. *Mucronea*, comprising two species of California, and *Centrostegia*, four species of the southwestern United States and northwestern Mexico, are clearly related to *Chorizanthe*. The ditypic *Lastarriaea* (Reveal, 1977), which is somewhat intermediate between these two complexes, occurs from the southern California Floristic Province to adjacent Baja California, and in Chile.

In *Eriogonum*, 113 of the approximately 247 species occur in California (Reveal and Munz, 1968; Reveal, 1970, 1971, 1972, 1977), 46 of them endemic. Of these, 69 occur in the California Floristic Province, with six of them only in northwestern Baja California (R. Moran, personal communication), and 44 of them are endemic to the California Floristic Province. Of the eight subgenera of *Eriogonum* (Reveal, 1969), two predominantly consist of annual species. Subg. *Ganysma* has 30 of its 59 species in California, including 5 annual species endemic to the California Floristic Province, and subg. *Oregonium*, which consists entirely of annuals, has 24 species in California and only six not found in the State. Of the total, 15 are endemic to the California Floristic Province, with a major center of speciation in southern California (Reveal, 1969). Three of the other six, entirely perennial, subgenera of *Eriogonum* are found in California, with the monotypic subg. *Clas-*

tomyelon endemic to limestone cliffs in the Death Valley region. As in the other genera discussed, most species of *Eriogonum* occur on the deserts outside of the California Floristic Province. The only genus with more species than *Eriogonum* in California is the Arcto-Tertiary genus *Carex*, with 143 native species, but only 22 endemics. The same sort of relationship is characteristic of *Chorizanthe*: of the 38 Californian species, 35 are found in the California Floristic Province, with five of them only in northwestern Baja California (R. Moran, personal communication), and 34 of these 40 are endemic; all are annual.

Summarizing for Eriogonoideae, there are 14 genera and about 321 species, with 12 genera and 159 species in California, and nine genera and 114 species in the California Floristic Province. The monotypic annual genus *Hollisteria* is endemic to the California Floristic Province, and *Lastarriaea* is nearly endemic, but has a second species in Chile. Of these 114 species, 80 (70%) are endemic, with 62 of the endemics being annuals. Despite the presence of the only perennial species of *Chorizanthe* in temperate South America, it appears that Eriogonoideae evolved in the semiarid Madrean region of western North America, probably in Oligocene time (see Reveal, 1977, for a discussion of the fundamental distinctiveness of the group). Their radiation into areas of mediterranean climate in California and explosive evolution there probably have taken place since the mid-Pliocene. This trend, which has affected mainly *Chorizanthe* and *Eriogonum*, has constituted a major event in the evolution of the family as a whole. The single South American species of *Oxytheca* and *Lastarriaea*, and the one South American annual species of *Chorizanthe*, all of which are endemic, probably reached that continent in the Late Pleistocene. *Sanmartinia*, a monotypic, annual genus described from South America, is reduced by Reveal (Reveal and Howell, 1976; Reveal, 1977) to *Eriogonum divaricatum* Hook. of Utah, Wyoming, Colorado, Arizona, and New Mexico and considered a recent introduction. The perennial species of *Chorizanthe* probably reached South America prior to the Pleistocene, and the South American species seem to have retained many primitive features. The ultimate source of Polygonaceae appears to have been tropical America (South America). Erigonoideae have radiated principally in the Madrean vegetation of North America and only secondarily in the geologically recent mediterranean climate of California.

CENTERS OF ENDEMISM IN THE CALIFORNIA FLORISTIC PROVINCE

Stebbins and Major (1965) provided a valuable discussion of endemism in California. Following Favarger and Contandriopoulos (1961), they recognize four classes of endemic species: (1) paleoendemics; (2) schizoendemics; (3) patroendemics; and (4) apoendemics. Paleoendemics are systematically isolated taxa, usually monotypic sections, subgenera, or genera, such as *Sequoia sempervirens* (D. Don) Endl., *Phacelia dalesiana* J. T. Howell (Constance, 1952), or *Eriogonum intrafractum* Cov. & Mart. Schizoendemics are vicarious endemics that have more or less simultaneously diverged from a common ancestor, such as the species of *Ceanothus* sect. *Cerastes.* Patroendemics are diploid or lower polyploid endemics that have given rise to more widespread higher polyploids; thus *Poa tenerrima* Scribn. ($n = 21$) is a highly restricted species related to the widespread *P. scabrella* (Thurb.) Benth. ($n = 42$); additional examples are given by Stebbins and Major (1965, p. 17-18). Finally, apoendemics are derivatives of polyploid parents; thus *Camissonia hardhamiae* Raven ($n = 21$) of southernmost Monterey and northern San Luis Obispo County is derived from the more widespread *C. micrantha* (Hornem. ex Spreng.) Raven (*n* = 7) and *C. intermedia* Raven (*n* = 14; Raven, 1969). Many additional examples from the flora of California were given by Stebbins and Major (1965, p. 20-22).

Intermediate habitats in general are richest in neoendemics (Stebbins and Major, 1965; Lewis, 1972; Stebbins, 1974, Chapter 8). In order to discover where neoendemism in the State is most pronounced, Stebbins and Major (1965) sampled 70 large genera of the California flora. The total numbers of endemic species in these genera show that the areas richest in these endemics are southern California, the Sierra, and the Central Coast. The poorest are the Central Valley and deserts. A high degree of endemism is found in areas which have a great variety of habitats and which during the Pleistocene were not subjected to severe climatic changes. Low mountains are associated with more endemism than very high mountains.

Stebbins and Major (1965, p. 16) found that patroendemics were especially frequent in the Central Coast region, in coastal or maritime sites, and at middle elevations in the Sierra Nevada. Apoendemics, on the other hand, appeared to be less closely confined to coastal areas, and were frequent in the inner Coast Ranges and other areas of recent habitats, such as higher parts of the Sierra Nevada.

Relict Areas

Stebbins and Major (1965) divide the state of California into ten floristic subdivisions (fig. 7). They attempt to assess the distribution of relict species within the State, defining these as all monotypic or ditypic genera confined to California or to a part of California and a neighboring region, as well as all species or species-pairs which are the only representatives of their genera in California and are separated from other species of their genera by a considerable distance. The distribution of these relict species shows two areas of maximum concentration (fig. 8). One is the Siskiyou-Trinity mountain area of northern California, where Arcto-Tertiary relicts predominate. The other is along the northern and eastern margin of the Colorado Desert, from the Little San Bernardino Mountains along the east slope of the San Jacinto and Santa Rosa Mountains, the Borrego Valley area, and southward into Baja California, where Madro-Tertiary relicts abound. As pointed out by Axelrod (1966b) these two regions represent areas in which relict species and genera associated respectively with the Arcto-Tertiary and Madro-Tertiary geofloras have persisted. It has been the interplay between these elements from the Miocene onward that has given the flora of California its richness. Between them, an area of rather low frequency of relict species compared with the diversity of the floras as a whole extends from the inner South Coast Ranges to the Tehachapi Mountains, an area of active speciation but not of relict endemism.

A correlation clearly appears to exist between areas of comparatively high precipitation in the warm season (April through September; fig. 9) and patterns of relict endemism and discontinuous plant distribution in California. The isohyets were drawn from data for numerous precipitation stations, supplemented by inferences of rainfall made from streamflow records in areas where precipitation records were few or lacking (Anonymous, 1941).

The areas of high warm season rainfall are mountainous, but they receive moisture from different sources. The northern half of the state is under the influence of cyclonic disturbances from the Gulf of Alaska. Since their effect decreases southwards, northwestern California has higher rainfall as well as a longer precipitation season. The second source lies

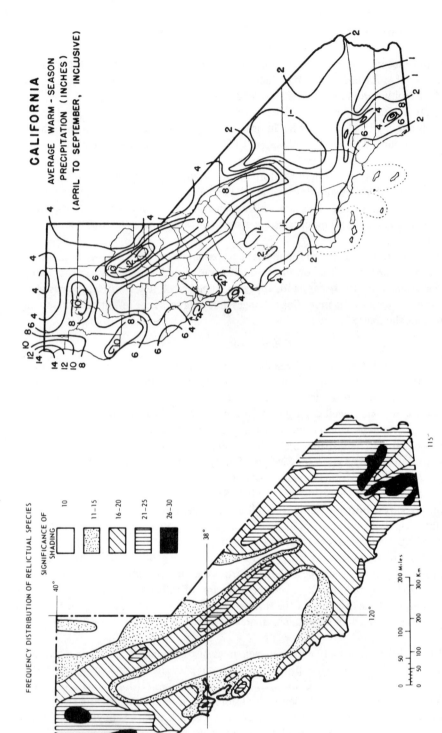

Fig. 8. Distribution of relict species in California. The key indicates that the black areas contain 26-30% of all relicts within the State, whereas the unshaded areas have 10% or less of all relicts. From Stebbins and Major (1965). The centers of high concentration of relict taxa coincide largely with areas of high warm-season precipitation (see fig. 9).

Fig. 9. Distribution of warm season (April-September) precipitation in California. From USDA Yearbook, Climate and Man (Anonymous, 1941). Compare with fig. 8.

115''

to the south, where the incidence of summer tropical storms over the mountainous areas gradually increases, adding to the duration and total of warm season rainfall in the mountains there. Six major areas of high endemism or of discontinuous distribution appear to coincide closely with the occurrence of relatively high warm season precipitation.

1. The Klamath-Siskiyou region of northwestern California and adjacent Oregon is well known for its numerous local endemics (Jepson, 1925; Whittaker, 1960; Stebbins and Major, 1965). It is also an area in which Tertiary relicts have survived, notably *Picea breweriana, Quercus sadleriana,* and others discussed above. The region also has one of the highest concentrations of conifers known anywhere, with 16 to 18 species occurring in proximity in several areas. Some of these conifers are typical of the Rocky Mountain flora (e.g., *Abies lasiocarpa, Picea engelmannii*), others are southern outposts of the Cascade flora [e.g., *Abies amabilis* (Dougl.) Forbes, *A. procera* Rehd., *Chamaecyparis nootkatensis* (D. Don) Spach] . Their persistence in this region apparently can be attributed to the local climate, which is more like that of the Late Cenozoic than anywhere else in the region (Axelrod, 1976b). The area has 15-20 cm precipitation in the warm season, a longer rainy season, and lower summer temperatures than occur inland or to the south in the Coast Ranges and Sierra Nevada, all factors which would tend to favor establishment and hence survival there.

2. The rich Coast forest of the Pacific Northwest ranges southward to the central California coastal strip (Clements, 1920). Three of the regular co-dominants, *Abies grandis, Picea sitchensis,* and *Tsuga heterophylla,* are rare in the forest south of Mendocino County, where warm season precipitation rapidly falls off as the Coast Range decreases in altitude (see fig. 9). A number of their regular associates have their southern limits a short distance southward in Marin County and in the Santa Cruz Mountains, where warm season precipitation rises again. Howell (1970) states that 84 taxa reach their southern limits in Marin County, largely in the moister highlands of Mt. Tamalpais and Point Reyes Peninsula. Thomas (1961) lists over 200 taxa that have their southern limits in the Santa Cruz Mountains. Most are herbs, but a number of woody plants are included, notably *Corylus cornuta* Marsh. var. *californica* (A. DC.) Sharp, *Ledum glandulosum* Nutt., *Quercus garryana* Dougl., *Rhododendron occidentale* (T. & G.) A. Gray, *Rubus leucodermis* Dougl. ex T. & G., *R. spectabilis* Pursh, and *Torreya californica* Torr. Some taxa in the Santa Cruz Mountains are disjunct from areas well north of Sonoma County where they are associates of the Coast forest.

In addition to these, some 146 taxa reach their southern limit in Monterey County (Howitt and Howell, 1964), including such species as *Botrychium multifidum* (Gmel.) Rupr., *Chimaphila menziesii* (R. Br.) Spreng., *Eleocharis pauciflora* (Lightf.) Link, *Garrya fremontii* Torr., *Montia parvifolia* (Moq.) Greene, *Polystichum dudleyi* Maxon, *Prunus subcordata* Benth., *Salix scouleriana* Barr., *Sequoia sempervirens* (Lamb.) Endl., *Trillium chloropetalum* (Torr.) Howell, *T. ovatum* Pursh, *Vancouveria planipetala* Call., *Viola sempervirens* Greene, *Whipplea modesta* Torr., and *Xerophyllum tenax* (Pursh) Nutt. At the same time, about 156 taxa reach their northern limit in Monterey County, mainly in the interior of the Santa Lucia Mountains, uplifted in the Late Pliocene-Pleistocene, and in the ranges to the east. Some 154 additional taxa reach their southern limit in San Luis Obispo County (177 their northern limit), among them *Hierochloë occidentalis* Buckl., *Nuphar polysepalum* Engelm., *Ribes glutinosum* Benth., *Salix sitchensis* Sanson, and *Trientalis latifolia* Hook. (Hoover, 1970). Finally, to mention a few similar examples of species that range farther south, *Ceanothus thrysiflorus* Eschs., *Gaultheria shallon* Pursh, *Oemleria cerasformis* (Hook. & Arn.) Landon (*Osmaronia*), *Pseudotsuga menziesii*, and *Vicia gigantea* Hook. reach Santa Barbara County (Smith, 1976); *Lithocarpus densiflorus* Ventura County, and *Myrica californica* the Santa Monica Mountains; other species reach the mountains of San Diego County and the California islands, as will be discussed below.

That members of the present Coast forest ranged south of their present limits in the recent past is apparent not only from range disjunctions, but also from the fossil record. *Picea sitchensis* is recorded at Tomales Bay (Mason, 1934) in the Millerton Formation, which has a radiocarbon age of 28,000 yrs B.P. (Axelrod, 1971). In addition, the Late Pliocene Sonoma flora (3.5 m.y.) from Santa Rosa has species inseparable from *Abies grandis, Chamaecyparis lawsoniana,* and *Tsuga heterophylla,* as well as taxa that are their regular associates on the Mendocino coast, notably *Alnus oregona* Nutt., *Chrysolepis chrysophylla* (Dougl. ex Hook.) Hjelmquist, *Myrica californica, Rhododendron macrophyllum* D. Don, and *Vaccinium parvifolium* Sm. (Axelrod, 1944).

3. The area of the northern Sierra Nevada, centering in the drainage of the Feather and Yuba rivers, has numerous taxa that otherwise occur chiefly in the moister parts of the Coast Ranges and Klamath Mountain region. Among the conspicuous taxa of this alliance are *Acer circinatum, Arbutus menziesii, Gaultheria ovatifolia* A. Gray, *Lithocarpus densiflorus, Rhamnus purshiana* DC., *Taxus brevifolia* Nutt., as well as numerous herbaceous plants. Few of them extend south of the latitude of Yosemite, where warm season precipitation rapidly decreases (fig. 9).

4. The high southern Sierra Nevada from Kearsarge Pass southward to Monache Mountain on the Kern Plateau is known especially for *Pinus balfouriana* Grev. & Balf., disjunct there from the Klamath Mountain region. Jepson (1925, p. 11) also noted that a number of other taxa are discontinuous only between these regions.

The disjunction seems explicable on the basis of a wider distribution of subalpine forest in the Quaternary, and elimination of taxa in the intervening parts of the Sierra where warm season precipitation is lower (fig. 9). Contributing factors to the present discontinuity in distribution probably are (a) the smaller area of the subalpine zone as the altitude of the range decreases northward, and (b) the impact of warm dry climate of the Xerothermic period which greatly affected the flora there. This is apparent from the occurrence of stands of *Pinus sabiniana* well up within the mixed conifer forest zone (e.g., Hetch Hetchy; N. of Hat Creek Ranger Station), and of *Pinus monophylla* Torr. & Frém. at similar or higher stations (e.g., upper Kings River basin, near Susanville [Sudworth, 1908]); and at Carson Pass (Taylor, 1976). Taylor (1976) notes that apart from *P. monophylla,* a number of other typically Great Basin taxa, such as *Betula occidentalis* Hook., *Cercocarpus ledifolius* Nutt., *Leucophysalis nana* (A. Gray) Averett and *Tetradymia canescens* DC. occur in the high mountains of Alpine County. He attributes their occurrence there to invasion during the warmer and drier Xerothermic (also see Axelrod, 1966b). The widespread andesite mudflow breccias would have provided especially favorable dry sites for their penetration far upward into the Range at this time. In addition, some typically Great Basin plants, including *Artemisia arbuscula* Nutt., *Cercocarpus ledifolius, Purshia tridentata* (Pursh) DC., and *Symphoricarpos vaccinioides* Rydb., probably migrated into the North Coast Ranges during the Xerothermic period and have persisted there (Taylor, 1976, p. 307).

5. A number of plants in the Peninsular Ranges of southern California are typical of the Great Basin and the central to southern Rocky Mountains. These notably include *Arctostaphylos patula* Greene var. *platyphylla* Wells [= *A. parryana* Lemmon var. *pinetorum* (Roll.) Wies. & Schrieb.] , *Pinus flexilis* James, and *Populus angustifolia* James. Additional genera which include species with this sort of distribution are listed in table 8. Some have stations in the New York, Clark, and Kingston mountains on the California-southern Nevada border (Henrickson and Prigge, 1975), and especially in the Charleston Mountains of southern Nevada (Clokey, 1951), where more numerous members of this alliance occur, as discussed recently by Axelrod (1976b: 32-33). Significantly, up to 20% of the total rainfall occurs there during the 3 summer months.

Many of these taxa were more prominent in western Nevada and California into the Pliocene, disappearing there as summer rains gradually decreased (Axelrod, 1973). Their persistence in the Peninsular Ranges of southern California, and locally on the east slope of the southern Sierra (e.g., *Pinus flexilis, Populus angustifolia*), also seems related to the incidence of considerable warm season precipitation there (fig. 9), a regular feature of the region of their optimum development.

6. Munz (1935, p. xxiv) detailed in a lucid manner the way in which taxa from central to northern California have penetrated in varying degrees into southern California. Of special significance is the large number of northern plants in the mountains of San Diego County that do not occur elsewhere in southern California, being disjunct there from areas largely north of San Luis Obispo County. He hypothesized they probably are "a relic of a once widespread flora; their persistence there may be explained by the rather high rainfall and rather mesophytic conditions of parts of the interior. . . ." Among these species are *Garrya fremontii, Quercus kelloggii* Newb., *Rosa gymnocarpa* Nutt. ex T. & G., *Rubus parviflorus, Salvia sonomensis* Greene, *Sedum spathulifolium* Hook. (Gordon and Grayum, 1976; F. Sproul, in prep.), *Vaccinium ovatum, Viola lobata* Benth., and *Zigadenus venenosus* S. Wats. (to northwestern Baja Calif.). *Limnanthes gracilis* Howell is disjunct between the Cuyamaca and Laguna Mountains of San Diego County and the Klamath Mountain region of southwestern Oregon, the only two regions where it occurs!

These taxa are clearly part of a larger group that ranged southward into the region during the moister, cooler parts of the Quaternary. This is consistent with distributional evidence elsewhere in southern California summarized by Munz (1935), and is suggested also by the occurrence of *Sequoia* in the Carpinteria flora near Santa Barbara, dated as older than 35,000 radiocarbon years (Axelrod, 1967a, 1967b). Additional evidence is provided by the Early Pleistocene Soboba flora ($\sim$ 1.5 m.y.) which shows that a mixed conifer forest, including *Abies concolor, Calocedrus decurrens* (Torr.) Florin, *Pinus lambertiana* Dougl., *P. ponderosa*, and *Populus tremuloides*, lived at the floor of San Jacinto Valley fully 1,000 m below the present lower limits of this forest in the nearby mountains (Axelrod, 1966b). Presumably the disjunct northern taxa in San Diego County and elsewhere in southern California disappeared from the intervening areas during the warmer and drier climate of the Xerothermic period (Axelrod, 1966b).

The preceding examples all occur in areas where high terrain naturally results in increased precipitation. However, terrain or precipitation in themselves probably are not the sole factors determining these distributions. If so, then the taxa should be more widely distributed in mountainous areas wherever rainfall is sufficiently high. That adequate warm season precipitation may be the decisive factor in their presence is inferred from the trend of climatic change to which they have adapted recently. Since summer rainfall decreased gradually over the region from the Late Miocene into the Early Quaternary, the appearance of summer-dry mediterranean climate is a very recent event, certainly during the past million years (Axelrod, 1971, 1973, 1976a, b). Thus the occurrence of even light warm-season rain may still have a critical role in seedling establishment, and hence in de- termining persistence in relict areas (Klamath-Siskiyou region) and in discontinuous ones (e.g., Cuyamaca Mountains).

Warm-season precipitation presumably ensures the establishment of seedlings, many of which are very susceptible to drought in the warm season, as reviewed by Fowells (1965) for many trees. In this regard, it is noteworthy that eight forest trees in the Sierra Nevada (i.e., *Abies magnifica* A. Murr., *Juniperus occidentalis* Hook., *Pinus albicaulis* Engelm., *P. contorta* var. *latifolia*, *P. flexilis* James, *P. monticola*, *Populus tremuloides*, *Sequoiaden- dron giganteum* [Lindl.] Buchh.) find their southern limits in Tulare County, precisely where total warm season precipitation rapidly decreases (fig. 9). It seems equally signifi- cant that the four of these species that reappear in southern California (*Juniperus occiden- talis, Pinus contorta* var. *latifolia, P. flexilis, Populus tremuloides*) are local there, and oc- cur chiefly in local areas with high warm-season precipitation. *Pinus contorta* var. *latifolia* and *Populus tremuloides* are abundant at the higher elevations of the Sierra San Pedro Mártir in northwestern Baja California, where late summer rains are characteristic. Other woody plants that have a similar distribution include *Acer glabrum* Torr., *Cornus nuttallii* Aud., *Euonymus occidentalis* Nutt. ex Torr., *Phyllodoce breweri* (A. Gray) Heller, *Rhodo- dendron occidentale*, and many herbaceous species as well (see lists in Munz, 1935).

These distributions in areas with high warm-season precipitation correspond closely to the principal relict endemic areas outlined by Stebbins and Major (1965; fig. 2), and re- produced here as fig. 8. The relation suggests that numerous distinctive genera and species in these endemic areas may be paleoendemics whose persistence owes chiefly to the high warm-season precipitation in these mountainous tracts. Autecological studies of selected taxa may provide further evidence concerning the validity of this theory.

Regions of Endemism

We shall now discuss in detail the patterns of endemism in the Central Coast Ranges, the Sierra Nevada, the California Islands, northwestern Baja California, and the Central Valley.

Central Coast Ranges.—Stebbins and Major (1965, p. 23, fig. 5) have analyzed local centers of endemism in the Central Coast Ranges, from southern Mendocino and Lake counties south to San Luis Obispo County. Such areas had been discussed early, especially by Jepson (1925) and by Mason (1946b) and more recently by Griffin (1975). The whole region is an active center of evolution, and comprises at least ten localized areas, each of which contains a relatively high number of endemic species (fig. 10).

Each of these areas was delimited to form an ellipse of approximately 500-800 km^2. As pointed out by Stebbins and Major (1965), all are mountainous and have a high local diversity of soil types and climate. The high numbers of endemic species in these areas are therefore thought to be associated with their high local diversity and the consequent opportunities for speciation. Relict species are most frequent in the areas with more moisture, and recently evolved species are most frequent in the drier areas. In terms of

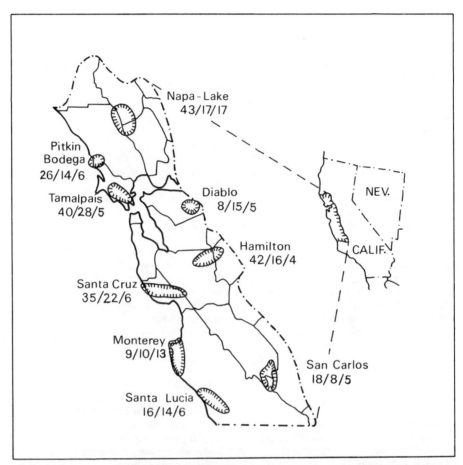

Fig. 10. Local endemic areas of the central Coast Ranges, California (simplified from Stebbins and Major, 1965, fig. 5). The numbers represent, respectively, the total endemic species in the intermediate plus large-sized genera, the number of relict species, and the species endemic to each particular area.

total endemism, the Mt. Hamilton and Napa-Lake areas stand highest on the list, with the Monterey and Tamalpais areas next. The proportion of endemics restricted to one of the areas, however, is highest in the regions with intermediate climate, which have presumably received species from various sources. In this respect, they resemble on a small scale the California Floristic Province as a whole, which combines a large number of persistent relicts with many recently evolved species to form a very rich local flora.

Sierra Nevada.—The middle elevations of the Sierra Nevada (ca. 500-2000 m) constitute important sites for the persistence of relict species as well as for the evolution of new endemic species. From the *Abies magnifica* and *Pinus contorta* var. *latifolia* forests at approximately 2000 m elevation and higher, there are fewer endemics. Subalpine habitats in the Sierra Nevada have come into existence only in the past few million years with the final uplift of the range. Those species that are endemic to subalpine and alpine habitats are therefore of particular interest (table 12). The derivation of this flora was considered

TABLE 12

Endemics of the Alpine and Subalpine Regions of the Sierra Nevada

Abronia alpina T. S. Brandegee	*Haplopappus eximius* H. M. Hall
Agropyron pringlei (Scribn. & Sm.) Hitchc.	*Haplopappus peirsonii* (Keck) J. T. Howell
Allium obtusum Lemmon	*Heuchera caespitosa* Eastw.
Aquilegia pubescens Cov.	*Hulsea brevifolia* A. Gray
Arabis pygmaea Roll.	*Ivesia muirii* A. Gray
Aster peirsonii C. S. Sharsm.	*Lewisia disepala* Rydb.
Astragalus austiniae A. Gray	*Lewisia sierrae* Ferris
Astragalus bolanderi A. Gray	*Lilium parvum* Kell.
Astragalus ravenii Barneby	*Linanthus oblanceolatus* (Brand) Eastw. ex Jeps.
Carex congdonii Bailey	*Lomatium torreyi* (Coult. & Rose) Coult. & Rose
Carex davyi Mkze.	*Lupinus covillei* Greene
Carex paucifructa Mkze.	*Lupinus hypolasius* Greene
Carex tahoensis Smiley	*Lupinus lobbii* A. Gray ex Greene
Castilleja culbertsonii Greene	*Luzula orestera* C. W. Sharsm.
Castilleja peirsonii Eastw.	*Mimulus barbatus* Greene
Castilleja praeterita Heckard & Bacigalupi	*Mimulus leptaleus* A. Gray
Chaenactis alpigena (A. Gray) M. E. Jones	*Mimulus whitneyi* A. Gray
Cirsium tioganum Congd.	*Nemacladus twisselmannii* J. T. Howell
Claytonia nevadensis S. Wats.	*Oreonana clementis* (Jones) Jeps.
Cryptantha crymophila I. M. Johnst.	*Orochaenactis thysanocarpha* (A. Gray) Cov.
Cryptantha humilis (Greene) Pays	*Oryzopsis kingii* (Bol.) Beal
Cryptantha nubigena (Greene) Pays	*Phacelia orogenes* Brand.
Delphinium polycladon Eastw.	*Phalacroseris bolanderi* A. Gray
Delphinium pratense Eastw.	*Phlox dispersa* C. W. Sharsm.
Dodecatheon subalpinum Eastw.	*Poa hansenii* Scribn.
Draba asterophora Pays.	*Polemonium eximium* Greene
Draba cruciata Pays.	*Primula suffrutescens* A. Gray
Draba sierrae C. W. Sharsm.	*Puccinellia californica* (Beetle) Munz
Erigeron miser A. Gray	*Saxifraga bryophora* A. Gray
Eriogonum breedlovei (J. T. Howell) Reveal	*Scirpus clementis* M. E. Jones
Eriogonum incanum Torr. & Gray	*Silene invisa* Hitchc. & Maguire
Eriogonum polypodum Small	*Silene sargentii* S. Wats.
Eriophyllum nubigenum Greene	*Streptanthus gracilis* Eastw.
Hackelia sharsmithii I. M. Johnst.	

in some detail by Smiley (1921), Sharsmith (1940), and Chabot and Billings (1972). The 68 endemics of this region (table 12) are mostly of rather recent origin because all but a few are closely related to other species found in adjacent parts of the Sierra Nevada. Those which do not have close relatives nearby seem to have differentiated from species that invaded the Sierra Nevada from the north during the Pleistocene.

Among the species that do not have close relatives nearby are *Aquilegia pubescens*, *Oreonana clementis*, *Polemonium eximium*, *Primula suffrutescens*, *Oryzopsis kingii* (Kam and Maze, 1974; Williams, 1975), and *Scirpus clementis*, the only taxa of their respective groups in the Sierra Nevada. They must be schizoendemics which diverged in the range following Pleistocene separation from relatives elsewhere (e.g., Chase and Raven, 1975). The genus *Phalacroseris* is endemic to subalpine habitats (as in *Abies magnifica* forest in the northern part of the range), but is one of a group, the Microseridinae (Feuer and Tomb, 1977), which radiated from Arcto-Tertiary habitats (table 4). A few very distinctive species—*Abronia alpina* (Wilson, 1970), *Arabis pygmaea*, *Castilleja praeterita* (Heckard and Bacigalupi, 1970), *Streptanthus gracilis, Nemacladus twisselmannii*—and the genus *Orochaenactis* are endemic to light sandy soils mostly above timberline in the southern-most Sierra Nevada; and the *Abronia, Nemacladus,* and *Orochaenactis*, at least, appear to be derived from ancestors that occurred in semi-desert, or at least very dry, habitats.

Major and Bamberg (1963, 1967) attempted to account for a small group of Rocky Mountain plants that occur together on limestone in the drainage of Convict Creek on the east slope of the Sierra Nevada in Mono County by migration across the Great Basin during the Pleistocene. As Chabot and Billings (1972, p. 174) have pointed out, however, there is so little similarity between the flora of the Sierra Nevada at large and that of either the northern or southern mountains of the Great Basin that such a path of migration appears highly unlikely. The decreasing proportion of Rocky Mountain species southward in the Sierra Nevada, coupled with the increasing proportion of endemism (Stebbins and Major, 1965), accords much better with a pattern of migration from the north. Additional reasons which add to the improbability of Pleistocene migration across the Great Basin are presented by Axelrod (1976b, p. 34-37).

Aside from these, the subalpine and alpine flora of the Sierra Nevada consists of a mixture of elements that came from the north, undoubtedly during the Pleistocene, like other more widespread species such as *Draba fladnizensis* Wulf, *Oxyria digyna* (L.) Hill, *Potentilla fruticosa* L., *Sibbaldia procumbens* L., and *Trisetum spicatum* (L.) Richt., and others derived from the semiarid ranges of western North America and the Mexican highlands as emphasized by Went (1948). The Pleistocene connections with the White Mountains discussed by Lloyd and Mitchell (1973) were probably one factor in bringing in various taxa from the Great Basin, but others may have come from the south. Most of the endemics in the subalpine and alpine regions of the Sierra Nevada have evolved *in situ* from species of the main forest belt. This took place under the influence of isolation from other sources of subalpine and alpine species, and under a climate that became progressively more arid in the summer (Chabot and Billings, 1972, p. 174).

The mountains of southern California share many species with the Sierra Nevada, and the ditypic genus *Oreonana* is endemic to the two areas. Many northern species reach even the mountains of San Diego County (Munz, 1935, p. xx). In addition, there are at least 72 species (Munz, 1935, p. xxi-xxii) endemic to the mountains of southern California, including 15 endemic to the San Bernardino Mountains alone, 30 to the San Gabriel Mountains, and 6 common to both.

California Islands.—The flora of the California Islands was recently reviewed by Thorne (1969). There are 23 endemic taxa in the northern group of islands, including Santa Rosa, Santa Cruz, San Miguel, and Anacapa, 38 in the southern group, and 15 common to both groups, for a total of 76 endemic taxa on the islands (Raven, 1967). Like the southern California coast and the adjacent portions of Baja California, the California Islands as a whole provide a refuge for more northern species that extended farther south during the Pleistocene pluvial cycles (Raven, 1963). Even San Clemente Island and Guadalupe Island, the most southerly of the group, have a number of species of definite northern affinities, among them *Carex tumulicola* Mackenzie, *Delphinium variegatum* T. & G. subsp. *thornei* Munz (Munz, 1969), *Pinus radiata, Polypodium scouleri* Hook. & Grev., *Ribes sanguineum* Pursh, and *Triteleia guadalupensis* Lenz (Lenz, 1975, p. 227). During the Late Pleistocene, such species as *Pseudotsuga menziesii, Ceanothus thrysiflorus,* and *Garrya elliptica,* now restricted to the mainland farther north, occurred on Santa Cruz Island (Chaney and Mason, 1930; Axelrod, 1967a, 1967b), where one tree of *Arbutus menziesii* still persists. The very distinctive genus *Lyonothamnus*, which had a wide distribution in the southwestern United States in Neogene time, is restricted to the four larger islands of the California group at present (fig. 5).

The flora of Guadalupe Island consists of some 164 native species, of which 31, including two genera (*Baeriopsis* and the extinct *Hesperelaea*) are endemic (R. Moran, personal communication). Guadalupe Island is about 265 km off the mainland of Baja California and about 390 km due south of San Clemente Island, with which 81 of its 164 native species are shared. In addition, there are strong affinities between the vegetation of the upper elevations of Cedros Island with the California group (Moran, 1972). The most outstanding characteristics of the islands as a whole are the high degree of endemism (see Weissman and Rentz, 1976) and the large number of northern taxa that occur on the islands, a trend also manifested in other groups (e.g., birds; Johnson, 1972).

Northwestern Baja California.—One of the most critical regions within the California Floristic Province is the northwestern corner of Baja California, including the crest regions and western slopes of the Sierra Juárez and Sierra San Pedro Mártir and extending south to approximately El Rosario at about 30°N lat. (Howell, 1957). The region also includes Cerro San Miguel and Cerro Matomí, where many northern plants reach their southern limits, as well as Guadalupe Island. According to Reid Moran (personal communication), there are some 1322 native and 114 introduced plants in northwestern Baja California of which about 227 do not occur in the state of California and 107 are endemic. Genera endemic in the area are *Adenothamnus, Baeriopsis, Hesperelaea* (now extinct), and *Ophiocephalus*. *Ornithostaphylos* was recently subtracted from this list by its discovery just within San Diego County (Moran, 1973). *Abutilon, Achyronychia, Allionia,* *Amauria, *Baileya, Bernardia,* *Blepharoneuron, *Brahea (Erythea; Moore, 1973, 1975), *Buddleja, Dalea,* *Digitaria, *Dracocephalum, *Drymaria, *Dyssodia, Echinocereus, Eucnide,* *Evolvulus, *Fouquieria,* *Harfordia, *Hedeoma, Hyptis,* *Ipomoea, *Lophocereus, *Lycurus, *Machaerocereus, Menodora,* *Myrtillocactus, *Nicolletia, Oxybaphus,* *Pachycereus, *Pectis,* *Pennellia, Petalonyx, Portulaca, Sanvitalia,* *Talinum, *Thamnosma, Tragia,* *Vauquelinia, and *Washingtonia* occur as native plants within this region but not within the California portion of the California Floristic Province. Of these, the 17 marked with an asterisk do not occur within the State of California.

As noted earlier, many species typical of the California Floristic Province occur on the higher slopes of mountains farther south in Baja California. For example, on the Volcán las Tres Vírgenes, a mountain just under 2000 m high, Reid Moran (personal communication) has recorded such species as *Ceanothus oliganthus* Nutt., *Heuchera leptomeria* Greene var. *peninsularis* Rosend., Butt., & Lak., *Keckiella antirrhinoides* (Benth.) Straw, *Lonicera subspicata* H. & A. var. *denudata* Rehd., *Rhus ovata* S. Wats., and *Xylococcus bicolor* Nutt., together with many other taxa of desert-border mountains, at a locality some 370 km south of the southeastern limits of the California Floristic Province. Other mountains in the central peninsula have various combinations of northern species, doubtless all survivors from more continuous Pleistocene distribution patterns.

The transition region between the chaparral and desert in Baja California is a particularly critical one in which many isolated endemics of northern affinities can be found (Shreve, 1936; Raven and Mathias, 1960). Examples are *Salvia chionopeplica* Epl., a member of the mainly Californian sect. *Audibertia*, and *Sanicula deserticola* Bell, most closely related to the Pacific coastal *S. bipinnatifida* Dougl. but separated from the nearest populations of that species by some 100 km. Northwestern Baja California is also an important area of survival, with a mild, equable climate. In view of this, it is not surprising that species such as *Ptelea aptera* Parry (Bailey, 1962), *Aesculus parryi* A. Gray (Hardin, 1957), *Fraxinus trifoliolata* (H. & A.) Lewis & Epl. (Lewis and Epling, 1940), and *Rosa minutifolia* Engelm., among the least specialized in their respective genera, are endemic in this region. They are reminiscent of other groups such as *Dicentra* (Stern, 1961), which are widespread but in which the least specialized species occur in California. In addition, *Adenothamnus* is one of the least specialized members of Asteraceae, subtribe Madiinae, and *Hesperelaea*, which unfortunately is now extinct, is a unique evergreen monotype in Oleaceae.

There is a definite connection between the mainland of northwestern Baja California and the California islands, as shown by the presence in both areas of such species as *Salvia brandegei* Munz, known only from Santa Rosa Island and near Santo Tomás and the evergreen *Ribes viburnifolium* A. Gray, which occurs on Santa Catalina Island and the coast of northern Baja California and in San Clemento Canyon, San Diego (R. Moran, personal communication). The distribution of the relict *Pinus torreyana* Parry ex Carr., known only from Santa Rosa Island and the coast of San Diego County, is comparable; in this context, *Prunus lyonii* (Eastw.) Sarg., which occurs on the islands off the coast of southern California and also at a few localities in the deep arroyos of south-central Baja California (e.g., San Julio Canyon), might also be mentioned. These areas share a relatively equable, maritime climate in which survival is facilitated.

This transitional area between coastal sage and the Sonoran Desert, in which these unique taxa have survived, may well simulate the environment in which they originated. New evidence now supplied by the Mint Canyon flora of Late Miocene age (Axelrod, unpublished), indicates that semidesert environments of this sort were already in existence in the interior, east of the Peninsular Range axis. The restriction of these unusual taxa to the environment, and the presence of similar conditions in the Miocene, suggests a considerable antiquity for them.

It would be highly desirable to have a modern floristic account of Baja California, and such a work is now in the course of preparation by Ira L. Wiggins. Thanks to his tireless efforts, and those of Reid Moran, Annetta Carter, and others, rich materials for such a flora are available.

Central Valley.—Much of the Central Valley south of Sacramento was inundated into the earliest Pleistocene. The present flora, therefore, dates largely from Middle Pleistocene time. There are approximately 44 species of vascular plants more or less strictly endemic to this region (table 13). Among these are the endemic *Neostapfia* and *Lembertia*. Many other genera, such as *Amsinckia, Astragalus, Lepidium, Lupinus, Mimulus,* and *Plantago,* are common and well represented in the Central Valley but they do not have any strictly endemic species.

A few of the endemic species of the Central Valley appear to have been derived from ancestors that came from the desert east of the Sierra Nevada: *Atriplex, Eremalche,* and possibly *Chamaesyce* are the only examples. All other endemics with the exception of *Neostapfia, Orcuttia* (Crampton, 1959; Reeder, 1965), and *Bacopa* are members of groups better represented in the Madrean woodland flora of the low foothills and valleys of California, and probably invaded the Valley as the Late Pleistocene lakes were drying. Some

TABLE 13

Endemics in the Central Valley of California

Atriplex cordulata Jeps.	*Juncus leiospermus* F. J. Herm.
Atriplex fruticulosa Jeps.	*Lasthenia chrysantha* (Greene ex A. Gray) Greene
Atriplex tularensis Cov.	*Lasthenia ferrisiae* Ornduff
Atriplex vallicola Hoov.	*Lasthenia fremontii* (Torr. & Gray) Greene
Bacopa nobsiana Mason	*Lasthenia platycarpha* (A. Gray) Greene
Boisduvalia cleistogama Curran	*Layia leucopappa* Keck
Calycadenia fremontii A. Gray	*Legenere limosa* (Greene) McVaugh
Chamaesyce hooveri (Wheeler)	*Lembertia congdonii* (A. Gray) Greene
Cirsium crassicaule (Greene) Jeps.	*Lessingia virgata* A. Gray
Cirsium hydrophilum (Greene) Jeps.	*Lilaeopsis masonii* Mathias & Constance (1977)
Cordylanthus hispidus Penn	*Microseris campestris* Greene
Cordylanthus palmatus (Ferris) Macbr.	*Monardella leucocephala* A. Gray
Cryptantha hooveri I. M. Johnst.	*Navarretia heterandra* Mason
Dichelostemma lacuna-vernalis Lenz	*Neostapfia colusana* (Davy) Davy
Downingia bella Hoov.	*Orcuttia pilosa* Hoover
Downingia ornatissima Greene	*Orcuttia greenei* Vasey
Eremalche kernensis C. B. Wolf	*Plagiobothrys austiniae* (Greene) I. M. Johnst.
Eriastrum hooveri (Jeps.) Mason	*Plagiobothrys scriptus* (Greene) I. M. Johnst.
Eryngium racemosum Jeps.	*Pseudobahia bahiifolia* (Benth.) Rydb.
Grindelia paludosa Greene	*Pseudobahia peirsonii* Munz
Grindelia procera Greene	*Trichostema ovatum* Curran
Hemizonia pallida Keck	*Tropidocarpum capparideum* Greene

of the endemics and other plants of the Valley are quite tolerant of, or even restricted to, saline soils of post-glacial age. A special feature of the region is the formation of winter-filled vernal pools on hardpan soils, where many endemic plants and animals occur (Jain, 1976).

Despite the presence of many typically desert plants, such as *Opuntia, Larrea, Prosopis,* and *Ephedra,* in and around the San Joaquin Valley especially (table 8), they do not appear to have contributed significantly to the endemism of the region. It is of interest that Brandegee (1893, p. 169) speaks of "the invasion of the San Joaquin Valley by the plants of the Mohave Desert now in active progress." Unfortunately we do not know which desert plants she considered to be invading the Valley in the 1890's, but the arrival of most desert elements there in the Xerothermic period or more recently seems assured, as reviewed by Axelrod (1966, p. 46).

In his outstanding study of endemism in the Central Valley, Hoover (n.d.) points out that many of the endemics there are the most advanced species of their respective groups: for example, those in *Monardella* and *Lessingia.* They seem to have originated under conditions of extreme aridity in the Valley within the last 8,000 years. Unfortunately, the entire flora of the Central Valley is in acute danger of extinction, owing to the almost complete conversion of the valley floor to agriculture (Jain, 1976). Many of the endemic species are endangered or threatened at present (Ripley, 1975). *Atriplex tularensis, Calycadenia fremontii,* and *Monardella leucocephala* are thought to be extinct already, and *Cordylanthus palmatus* apparently became extinct in the wild only to be reintroduced near the stations where it once occurred (L. R. Heckard, personal communication). Much attention must be given to conservation if even a part of the fascinating flora of the Central Valley of California is to be saved for future generations to study.

Origins of the Endemic Plants

In the Late Pliocene, before the final elevation of the Sierra Nevada-Cascade axis, the Transverse Ranges and the Peninsular Ranges, and the development of a full-mediterranean climate in California, the present area of the California Floristic Province presumably was inhabited by about the same number of genera as at present, but fewer species—perhaps 3,400 instead of the 4,452 at present. Monocots probably comprised a quarter of the total number of species (18-19% at present), and the percentage of annuals was probably about 20% (28% at present). Not more than a third (instead of 48%) of the species probably were endemic. These estimates have been obtained by comparison with other regions with similar climate and topography.

Some outstanding examples of groups now strongly associated with, or well developed in, the area at present are: Asteraceae-Lactuceae and Madiinae, Hydrophyllaceae, Liliaceae-Calochorteae, Limnanthaceae, Onagraceae-Epilobieae and Onagreae, Papaveraceae-Eschscholzioideae and Platystemonoideae, Polemoniaceae-Gilieae, Polygonaceae-Eriogonoideae and Scrophulariaceae-Rhinantheae and perhaps Cheloneae. These groups can reasonably be inferred either to have originated in Madrean vegetation in southwestern North America or at best to have had their primary seat of differentiation there. They, as well as many other genera (table 2), were probably in the region by the Pliocene and provide the distinctive character for the flora of California at present.

With the episodes of mountain building, cooling, and drying of the climate that were accelerated in the Late Pliocene and have continued to the present, the numbers of species in many of these genera have been multiplied (Axelrod, 1967c; Stebbins, 1952, 1974). Especially among annual dicots of certain families there has been an impressive proliferation of species. The products of such proliferation, existing side-by-side with relicts surviving in local areas of equable climate, especially those in which a fair amount of precipitation falls in the summer, are a distinctive element in the flora of California, but not an ancient one. They have developed in relation to the ecotone between Arcto-Tertiary and Madro-Tertiary vegetation which was established in the Miocene, and multiplied greatly during Quaternary time, as discussed in detail by Axelrod (1966b, p. 55-60).

EDAPHIC ENDEMISM

The restriction of plants to certain soil types is an important feature of the flora of California, as it is in most arid or semiarid regions of the world. The kinds of patterns found in the State were discussed at some length by Mason (1946a, 1946b), and then studied experimentally by Kruckeberg (1951, 1954), Walker (1954), and others. In general, edaphic endemism is related not to a particular requirement for a certain kind of soil, but rather to the absence of competition on a soil type that is unique for the area where it occurs (Kruckeberg, 1951, 1954; Rune, 1954). Plants at the margins of their ranges notably tend to grow on soil types that are not occupied by the dominant vegetation of the particular area, a pattern that Raven (1964) suggested would lead to the origin of endemics restricted to unusual edaphic situations as the marginal populations diverged.

Plants are restricted to many kinds of substrate in California. Some occur on sands along the coast; examples are *Abronia latifolia* Eschs., *Ambrosia chamissonis* (Less.) Greene, *Camissonia cheiranthifolia* (Hornem. ex Spreng.) Raimann, *Carpobrotus aequilaterus* (Haw.) N. E. Br. [*C. chilensis* (Mol.) N. E. Br.], and *Cirsium rhothophilum* Blake. The beach vegetation of the Pacific Coast from Baja California north has recently been reviewed by Breckon and Barbour (1974). Such species as *Arctostaphylos silvicola* Jeps. & Wiesl. and *Erysimum teretifolium* Eastw. are restricted to Miocene marine sandstone in the Santa Cruz Mountains, where until recently an extra-limital population of the middle- to upper-elevation Sierran *Calyptridium monospermum* Greene (Hinton, 1975) also occurred.

A particularly interesting edaphically delimited region in coastal northern California is the "pygmy forest" of Mendocino County (McMillan, 1956, 1959, 1964; Jenny, Arkley, and Schultz, 1969). Here stunted trees of *Cupressus pygmaea* (Lemmon) Sarg., *Pinus muricata,* and *P. contorta* var. *bolanderi* (Parl.) Vasey occur within a forest belt consisting of normal-sized trees. They are confined to a podsol soil with a white, bleached surface horizon and an ironlike hardpan layer beneath. No species are known to be endemic to the "pygmy forest," but at least two geographical races, *Arctostaphylos nummularia* A. Gray var. *nummularia* and *Pinus contorta* var. *bolanderi*, have such a distribution (Kruckeberg, 1969). A few species of higher plants are restricted to deposits of seabird guano on islands (Ornduff, 1965). One which occurs in California is *Lasthenia minor* (DC.) Ornduff subsp. *maritima* (Gray) Ornduff, from the Farallon Islands northward.

Various kinds of volcanic substrates are likewise important in controlling plant distributions in California; such patterns have been discussed by Mason (1946b). As examples we mention *Arctostaphylos elegans* Jeps., *Ceanothus purpureus* Jeps., *Dudleya parva* Rose & Davids., *Eriastrum brandegeae* Mason, *Eriogonum crocatum* A. Davids., *Madia nutans* (Greene) Keck, and *Polygonum bidwelliae* S. Wats. Other special examples of edaphic endemism on nonvolcanic soils are *Monardella antonina* Hardham and *Eriogonum eastwoodianum* J. T. Howell on diatomaceous shales in the inner South Coast Ranges; and *Arctostaphylos myrtifolia* Parry, *Helianthemum suffrutescens* Schreib., and *Eriogonum apricum* J. T. Howell on Eocene laterite and sericitic schists in the Sierra Nevada foothills especially around Ione, Amador County (Gankin and Major, 1964).

Restriction to limestone soils is not as common in California as in the Mediterranean region, simply because limestone and other carbonates are not widespread rocks in California. However, they increase to the east, where they make up most of the ranges of central and eastern Great Basin. In that area there are some significant instances of taxa restricted to carbonates, as in the desert ranges of eastern California and adjacent Nevada. Thus *Pinus longaeva, Cercocarpus intricatus* S. Wats., *Petrophytum caespitosum* (Nutt.) Rydb., and *Erigeron clokeyi* Cronq., among other species, are almost entirely confined to soils derived from dolomite in the White Mountains of eastern California (Lloyd and Mitchell, 1973).[1] *Cheilanthes sinuata* (Lag.) Domin var. *cochisensis* (Goodd.) Munz, *C. jonesii* (Maxon) Munz, *C. feei* T. Moore, and *Adiantum capillus-veneris* L. are among the ferns that occur mainly or exclusively on limestone or other calcareous soils in California. In the mountains around Death Valley, as mentioned above, such restricted endemics as *Eriogonum intrafractum, Tetracoccus ilicifolius, Maurandya petrophila* Cov. & Mort., and *Phacelia mustelina* Cov. are restricted to limestone cliffs.

On these smooth-faced cliffs, plants growing in crevices may receive much more moisture than those growing under the same conditions of precipitation on nearby rough-surfaced flats and slopes (Danin, 1972; Danin et al., 1975). On the cliffs, the water may run down into the crevices, rather than percolating rapidly down into the soil. Thus, many of

---

1. See also D. E. Marchand, *Ecology* 54:233-250 (1973).

the plants endemic in or confined to the crevices of cliffs may be relicts that were more widespread in sclerophyllous vegetation under moister conditions during the Pleistocene. Despite the seeming great aridity of the cliffs, the plants that grow in their crevices may be more mesophytic in their requirements than those of the rough slopes. Furthermore, the shade in the cliffs provides protection from the high sun of summer and thus reduces evaporation, increases effective moisture, and hence ensures further their persistence in these restricted sites.

One of the most unusual assemblages of plants on limestone in California, that in the Convict Creek basin of Mono County on the east side of the Sierra Nevada, has been studied by Major and Bamberg (1963, 1967). Here occur together *Arctostaphylos uva-ursi* (L.) Spreng. at its only station in the Sierra, *Draba nivalis* Liljebl. var. *elongata* S. Wats., *Kobresia myosuroides* (Vill.) Fiori & Paol., *Salix brachycarpa* Nutt., and *Scirpus rollandii* Fern., the last four species not known elsewhere in California, but disjunct from the Rocky Mountains, the eastern Great Basin, and the Cascades. *Pedicularis crenulata* Benth. occurs at its only California station near the mouth of Convict Creek. As mentioned above, this unusual assemblage of plants probably came to the Sierra Nevada from the north and persisted on the unusual soils of the Convict Creek basin owing to a lack of competition with the dominant flora of the granitic rocks of most of the remainder of the region.

Serpentine Soils

Many species of vascular plants in California are wholly or partly restricted to serpentine and associated ultrabasic rocks in California. A number of papers have been devoted to them, some discussing the evolutionary processes that account for their origin (Stebbins, 1942a; Raven, 1964), some analyzing soil-plant relationships (e.g., Walker, 1954; Kruckeberg, 1954), and others considering the role of serpentine as an edaphic factor in plant distribution (e.g., Mason, 1946a, 1946b; Kruckeberg, 1954, 1969; Whittaker, 1960; Procter and Woodell, 1975). In his discussion of the serpentine problem, Kruckeberg (1954, 1969) suggested that adaptation to serpentine followed by biotype depletion and development of isolated populations into local endemics provides an explanation of serpentine endemism. This seems acceptable for endemism in local areas, but does not clarify either the problem of discontinuous distribution, or the reasons for biotype depletion in non-serpentine areas. Many taxa seem to have originated as serpentine endemics by saltational speciation (Lewis, 1962) in marginal populations (Raven, 1964). As we have seen, marginal populations often are on edaphic situations that are unique for the species as a whole.

On the other hand, such reasoning does not explain the manner by which such woody species as *Cupressus sargentii* Jeps. (fig. 11), *Quercus durata* Jeps., and other taxa attained their present disjunct distributions on ultrabasic rocks. Because the serpentine problem has received so much attention, we shall here discuss it in some detail. We propose that some of the present taxa had a wide occurrence on non-serpentine sites from the Miocene well into the later Pliocene, that they invaded serpentine sites as these became available in the Pliocene, and that they were later confined to serpentine areas as the more widespread ecotypes disappeared as summer rains decreased, and that adaptation to ultrabasic substrates removes them from competition with the adjacent, non-serpentine flora. Several factors have had a role in this inferred history.

Age.—Distinction must be made between the Jurassic and Early Cretaceous ages of serpentine and allied ultrabasics, which include both intrusive igneous and metamorphic

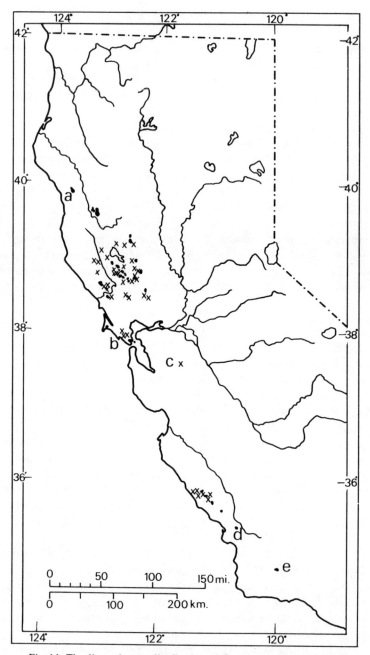

Fig. 11. The discontinuous distribution of *Cupressus sargentii* in the California Coast Ranges (from Griffin and Critchfield, 1972). Note especially the disjunct areas on serpentine at Red Mountain (a), Mt. Tamalpais (b), Cedar Mountain (c), near Cuesta Pass (d), near Zaca Peak (e). Solid black areas are stands more than 5 km. across; X indicates smaller stands or those of unknown size.

rocks, and the time when they were exposed and became available for plant occupation. The oldest serpentine areas are those that gave rise to the detrital serpentine deposits in the Early Cretaceous marine section of the North Coast Ranges. They are not of concern, however, because the taxa under consideration were not yet in existence. Similarly, areas of ultrabasic rocks in the Sierra Nevada were exposed in the Late Cretaceous to Paleocene, considerably before the appearance of taxa that are confined to serpentenite areas today. In any case, the ultrabasic rocks in both areas were then covered by diverse marine (in the Coast Ranges) and volcaniclastic (in the Sierra Nevada) Tertiary formations that were stripped subsequently to expose the present serpentine areas. To judge from evidence depicted on the Geologic Map of California (Anonymous, 1958-1968), the ages of the serpentine outcrops where taxa of discontinuous distribution occur are essentially as follows.

Mendocino County: Current marine microfossil evidence indicates that rocks previously mapped as Franciscan Formation in the coastal belt from the Fort Bragg-Willits area north to beyond Cape Mendocino-Scotia, are of Eocene, not Cretaceous age (Evitt and Pierce, 1975). The margin of the sea extended farther inland, as shown by the marine Paleocene and Eocene rocks near Round Valley, and Middle Miocene marine rocks are also in the area (Ukiah Sheet). Since the Pliocene Wildcat Formation to the northwest displays a similar degree of deformation, uplift and erosion to expose areas of ultrabasic rocks probably were accomplished chiefly in the later Pliocene and Early Quaternary.

Lake-Napa-Sonoma Counties: Over much of this region the serpentine areas were largely covered by Sonoma Volcanics, dated at 3.5 m.y., and also by younger volcanics in the Clear Lake region which are of Quaternary age.

Sierra Nevada: The linear, discontinuous belt of serpentine from Quincy southward was covered by Miocene and Pliocene andesite mudflow breccias and associated sedimentary rocks that were still being deposited into the Middle Pliocene ($\sim$ 6 m.y.). Hence, most areas were exposed only after uplift enabled erosion to strip the volcanic cover, largely during and after the Late Pliocene.

At the lower, northwest corner of the Sierra, ultrabasic rocks are exposed in the valleys of Big Chico Creek and Butte Creek. The adjacent interfluves are made up of the Tuscan Formation, dated at 3.2 m.y. Hence, those serpentine areas were exposed only after the present streams had trenched fully 300 m down through the Tuscan Formation, presumably during the Middle Quaternary.

Santa Lucia Range: The range was covered by seas into the Late Miocene ($\sim$ 10-12 m.y.). Since the present areas of ultrabasic rocks were exposed after the range was elevated and the marine cover stripped from it, they are largely Pliocene to Early Pleistocene.

San Rafael Mountains: The coastal section from San Luis Obispo southward to Figueroa Mountain was largely covered by seas in the Miocene, and locally (as in the Huasna Basin) into the Middle Pliocene. The areas of serpentine were exposed largely during the Pliocene and Early Quarternary.

Diablo-La Panza Ranges: These ranges were mobile in the Tertiary, for they border the San Andreas fault. Mt. Diablo is a diapir, and ultrabasic rocks were exposed there only in the Late Pliocene and more recently. The large serpentine area in the New Idria region is also a diapir. It may have been exposed in the Late Miocene, as suggested by serpentine detritus that characterizes the Big Blue Formation north of Coalinga. Hackel (1966) maintains that the Diablo uplift, from Coalinga northward, has been a positive area (with serpentine exposed) throughout the Tertiary, but his evidence is not convincing (C. Durrell, oral communication, December, 1975).

The preceding data indicate that the present discontinuous areas of ultrabasic rocks were exposed largely during the Late Pliocene and Early Quaternary as erosion stripped overlying sedimentary and volcanogenic rocks. Since many woody taxa that are now on serpentine were already in existence in Miocene time, they must have ranged widely on other, diverse rocks at that time, and have become adapted to ultrabasic substrates more recently.

Degree of Restriction.—Species that occur on serpentine and other ultrabasics differ in their degree of confinement to those rocks. These can be grouped into three assemblages. Group I includes species of woody plants that occur on ultrabasic rocks and are also widely distributed on other substrates, as

*Abies bracteata (D. Don) Nutt.
*Adenostoma fasciculatum
*Arbutus menziesii
*Calocedrus decurrens
*Ceanothus cuneatus (Hook.) Nutt.
*Cercocarpus betuloides Nutt. ex T. & G.
*Chamaecyparis lawsoniana
 Chrysolepis chrysophylla var. *minor*
*Dendromecon rigida Benth.
*Heteromeles arbutifolia (Lindl.) M. Roem.
*Holodiscus discolor (Pursh) Maxim.
 Lithocarpus densiflorus var.
  echinoides (R. Br.) Abrams

Pickeringia montana Nutt.
*Pinus attenuata Lemmon
 Pinus jeffreyi
*Pinus sabiniana
*Pseudotsuga menziesii
*Quercus agrifolia
*Quercus lobata Née
 Quercus vacciniifolia
*Rhamnus californica Eschs.
*Styrax officinalis L. var.
  californica (Torr.) Rehd.
*Umbellularia californica (H. & A.) Nutt.

Those marked by an asterisk (*) are represented in Miocene and Pliocene floras by essentially the same species, though the fossils were assigned different names (for reasons, see Chaney, 1948; Axelrod, 1967a). They occur in southern California, western Nevada, eastern Oregon, and Idaho, in areas well removed from ultrabasic rocks. This distribution assumes considerable significance inasmuch as ultrabasic rocks had only a very restricted distribution in California at that time. As for the few taxa (above) that do not have a fossil record, they no doubt had a wider distribution in the past because serpentine areas have only recently been uncovered. *Pickeringia* is a unique monotypic genus well removed from other alliances in the Fabaceae (Faboideae) and is inferred to have considerable antiquity, probably pre-Miocene at least. A similar age may be inferred for *Adenostoma*, a unique genus in Rosaceae that ranges widely on serpentine and other substrates as well. The others may be as old as Miocene, though definitive evidence is lacking. *Chrysolepis chrysophylla* and *Lithocarpus densiflorus* are both in the Miocene as ancestral forms of the shrub varieties, and the same applies to *Quercus chrysolepis* Liebm., the ancestor of the high montane *Q. vacciniifolia*. There is, therefore, reason to assert that all of these taxa probably became adapted to live on serpentine long after the Miocene.

Numerous herbaceous genera contain species that occur on serpentine and other ultrabasics but also range widely off of them. A number of genera such as *Bromus, Calystegia, Caulanthus, Dentaria, Eschscholzia, Gilia, Lasthenia, Lepidium, Linanthus, Microseris, Onychium, Orthocarpus, Sidalcea, Sitanion, Stipa,* and *Viola* are common on soils derived from serpentine rocks but, in our present state of knowledge, no species of these genera appear to be entirely confined to such soils. On the other hand, as noted by Kruckeberg (1969), many taxa are conspicuous by their absence or near absence on serpentine soils:

among them, Ericaceae, *Penstemon*, Ranunculaceae, Rosaceae, and Saxifragaceae, and, possibly by inhibition of the root-nodule-forming bacteria they require, Fabaceae. These taxa seem, for various reasons, to have been unable to occupy serpentine soils. A similar role for microorganisms in controlling the restriction of certain plants to serpentine is suggested by the results of Tadros (1957) with *Emmenanthe*. Local moist seeps at the base of slopes of ultrabasic rocks are the chief habitat for many of the endemics and other unusual herbaceous plants of the Siskiyou region: *Cypripedium californicum* A. Gray, *Darlingtonia californica* Torr., *Lilium bolanderi* S. Wats., *L. occidentale* Purdy, and *Trillium rivale* S. Wats., as noted by Kruckeberg (1969), although some extend into other areas such as coastal dunes, where there may also be a low calcium-magnesium ratio (Proctor and Woodell, 1975, p. 336).

Group II includes a number of woody taxa that have a discontinuous distribution on serpentine and are largely, but not wholly, confined to it. *Cupressus macnabiana* A. Murr. is a good example. As summarized by Griffin and Stone (1967), it occurs on rocks as diverse as Eocene sandstone near Montgomery Creek, rhyolite tuff in Sonoma County, Cretaceous sandstone in Yolo County, meta-rhyolite near Whiskeytown and Pleistocene basalt on Ash Creek and Lack Creek southeast of Redding, as well as on diverse outcrops of ultrabasic rocks. It is sufficiently different from other members of the genus in California to indicate that it may be Miocene, or older. *Quercus durata* is another example of a species that ranges widely on serpentine and related rocks, but occurs also on more normal substrates. It is locally on rhyolite tuff on the west slope of Howell Mountain, Napa County; on sites (unspecified) away from serpentine in San Luis Obispo County (Hoover, 1970); at sites in the San Gabriel Mountains, southern California, where it is on gneiss and schist, not ultrabasic rocks (McMinn, 1951, p. 83); in addition John M. Tucker reports (oral communication, July, 1975) that populations of *Q. dumosa* Nutt. in San Diego County have characters that indicate introgression with *Q. durata*, implying that it was once in that area, which is well removed from ultrabasic rocks.

There are additional taxa that appear to belong to this group. Uncertainty stems from the fact that data on herbarium sheets frequently do not indicate the edaphic occurrence of a taxon, and locality data often are not sufficiently precise to enable one to consult a geologic map so as to infer their edaphic occurrence. Examples of species that may belong to this group are *Garrya buxifolia* A. Gray and *Ceanothus pumilus* Greene in northwestern California and adjacent Oregon; *G. congdonii* Eastw. of the inner Coast Ranges and Sierra foothills; and *Salix breweri* Bebb and *Ceanothus jepsonii* Greene of the Coast Ranges. *Cupressus sargentii,* mentioned above, likewise appears to fall into this pattern, as does *Berberis pumila* Greene from the northern Sierra Nevada, the Klamath Mountain region, and the Siskiyou Mountains of southern Oregon. All are sufficiently distinct to suggest that they have some antiquity, possibly Miocene.

Group III taxa are chiefly herbaceous and are wholly confined to ultrabasic rocks, usually in restricted local areas. A number of such species are listed in table 14, but the list is very incomplete, for reasons discussed above, and omits *Hesperolinon, Navarretia,* and *Streptanthus,* discussed in the following section. Those species confined to a single locality or area on serpentine rock are marked with an asterisk in this table. Most of these species may have originated locally and in Late Pleistocene or Recent time.

Examples.—Three genera of herbaceous plants might be mentioned as examples of the interplay between evolutionary radiation and edaphic endemism in California. First, *Na-*

TABLE 14

Examples of Herbaceous Plants Nearly or Entirely Restricted to Serpentine and
Related Soils in California. Those known from a single locality are indicated by an asterisk.
Hesperolinon, Navarretia, and *Streptanthus* are discussed in the text.

Acanthomintha lanceolata Curran	*Claytonia gypsophiloides* F. & M.
Allium hoffmanii Ownbey	**Cordylanthus nidularius* J. T. Howell
Allium sanbornii Wood	*Cryptantha mariposae* I. M. Johnst.
Allium serratum S. Wats.	**Dudleya bettinae* Hoov.
Antirrhinum subcordatum A. Gray	**Eriogonum congdonii* (S. Stokes) Reveal
Aquilegia eximia Van Houtte ex Planch.	*Eriogonum hirtellum* Howell & Bacig.
*Arabis mcdonaldiana Eastw.	*Eriogonum tripodum* Greene
Arctostaphylos obispoensis Eastw.	*Fritillaria falcata* (Jeps.) D. E. Beetle
Arenaria howellii S. Wats.	*Fritillaria purdyi* Eastw.
Arnica cernua Howell	*Gutierrezia californica* (DC.) T. & G.
Asclepias solanoana Woodson	(Solbrig, 1965)
Brodiaea pallida Hoover	**Haplopappus ophitidis* (J. T. Howell) Keck
Brodiaea stellaris S. Wats.	**Layia discoidea* Keck
Calamagrostis ophitidis (J. T. Howell) Nygren	*Linanthus ambiguus* (Rattan) Greene
*Calochortus tiburonensis A. J. Hill (1973)	*Lomatium howellii* (Wats.) Jeps.
*Camissonia benitensis Raven	*Madia hallii* Keck
*Carex obispoensis Stacey	*Mimulus brachiatus* Penn.
*Castilleja neglecta Zeile	**Mimulus nudatus* Curran ex Greene
Cheilanthes carlotta-halliae Wagner & Gilbert	*Monardella benitensis* Hardham
Chlorogalum grandiflorum Hoov.	**Phacelia greenei* J. T. Howell
Cirsium campylon H. K. Sharsm.	*Senecio clevelandii* Greene
Cirsium fontinale (Greene) Jeps.	*Senecio ligulifolius* Greene
*Clarkia franciscana Lewis & Raven	*Zigadenus fontanus* Eastw.

varretia, a genus of some 29 species in North America, 24 endemic to and all found in the California Floristic Province, together with one species in South America (Crampton, 1954). Many of these occur in vernal pools on clay soil, but as Mason (1946b, p. 255) pointed out, others are highly restricted edaphic endemics: *N. pleiantha* Mason in an acid bog, in soil rich in organic materials; *N. pauciflora* Mason in volcanic ash heavily strewn with acidic obsidian rubble; *N. mitracarpa* Greene subsp. *jaredii* (Eastw.) Mason in serpentine and related clay soils; and *N. jepsonii* V. Bailey to vernal pools on serpentine. These examples could be multiplied, but serve to indicate the range of edaphic diversity in the genus and the importance of the edaphic factor in its evolution.

Streptanthus has long been known as a genus with many serpentine endemics in the California region. Some 24 of the 30 to 35 species of the genus occur in California, 22 of them (all endemic) in the California Floristic Province. At least 14 of these species occur on serpentine, of which 12—including seven of narrow range and five that are more widespread—are found nowhere else (Hoffman, 1952; Kruckeberg, 1969, p. 142). In addition, *Streptanthus glandulosus* Hook. is widespread on various soil types, but is often on serpentine, and *S. tortuosus* Kell. usually does not occur on serpentine but has some local races which are so restricted. The former has been studied in some detail by Kruckeberg (1951), who found that plants representing ecotypes that did not originate on serpentine soil were clearly intolerant of such soil, whereas indistinguishable serpentine samples grew vigorously on serpentine. Kruckeberg (1957, 1958) has also analyzed the fertility of hybrids between isolated populations of *S. glandulosus* and of the related more local species *S. albi-*

dus Greene and *S. niger* Greene. There was a significant negative correlation between the degree of hybrid fertility and the distance separating the two parents, which again suggests a pattern of increasing restriction to serpentine and biotype depletion. The whole complex might be taken as analogous to *Quercus durata*, discussed above. Increasing adaptation to serpentine soils need not necessarily be thought of as biotype depletion. The sorts of patterns that are observed would originate automatically if serpentine afforded one of the few available habitats for the group and the plants were able to grow there.

The most extreme of all serpentine genera, in a sense, is *Hesperolinon*. All 12 species are annuals, all occur on serpentine soil, and eight are entirely confined to it (Sharsmith, 1961). Two are highly restricted in range, but one, *H. disjunctum* Sharsm., although entirely restricted to serpentine soil, is disjunct between the inner North and South Coast ranges—a distance of some 130 km from Napa County to the Mt. Hamilton Range, then 150 km farther to southern San Benito County. *Salix breweri* and other species have similarly disjunct ranges, and *Astragalus clevelandii* Greene together with a few other species is disjunct between the inner North and South Coast ranges for a distance of nearly 300 km. It is difficult to imagine that the disjunction in range in the *Hesperolinon*, a specialized member of a rapidly evolving complex of annual herbs, took place as long ago as that which characterizes the other two very distinctive and probably relict woody species. Perhaps, therefore, the southern populations assigned to *H. disjunctum* were derived directly from *H. micranthum* (A. Gray) Sharsm., with which they occur, and are not in that region related to the northern ones (cf. Sharsmith, 1961, p. 293). At any event, *Hesperolinon* would be a most interesting subject for biosystematic study, as suggested by Sharsmith (1961).

Climate.—Commencing in the Miocene and continuing down through the Pliocene, there was a general trend toward greater ranges of temperature, and a major shift in the seasonal incidence of precipitation. From a Miocene climate with ample summer rain, warm season precipitation gradually decreased so that the wet season was concentrated progressively into the cooler half of the year (Axelrod, 1973, fig. 2). Many taxa that were present in the region were eliminated as summer rainfall decreased. Some are now in the eastern United States (*Carya, Diospyros, Fagus, Magnolia, Persea, Nyssa, Taxodium, Ulmus*), others are in eastern Asia (*Albizia, Ailanthus, Cercidiphyllum, Ginkgo, Metasequoia, Zelkova*), and some are represented now by taxa that are in the southwestern United States and Mexico (*Bumelia, Cedrela, Pithecellobium, Robinia, Sapindus*). In addition, ecotypes of numerous taxa became extinct, thus giving the surviving species a more restricted distribution than formerly (Axelrod, 1940, 1976b, fig. 4). The woody taxa that persisted were already adapted to live under the emerging mediterranean climate (Axelrod, 1973, 1975). In other words, survival involved a shift in the critical period of establishment into the earlier, moister half of the year, a response that no doubt was controlled genetically.

Evolution.—The varied range of tolerance of taxa with respect to ultrabasic rocks, as illustrated by the taxa in Groups I to III, points to the ways in which species restricted to ultrabasic rocks may originate. From the widely adapted species exemplified by Group I, there is a gradual progression to those in Group II that are more narrowly confined to serpentine. The eventual elimination of biotypes from non-serpentine areas would result in the unique endemics confined to serpentine. These are to be contrasted with the neoendemics of Group III, which are more typical of rapidly radiating groups in areas of high topographic-climatic-edaphic diversity. *Arctostaphylos* and *Ceanothus*, woody genera with a pattern of radiation similar to that of the herbaceous groups just mentioned, are discussed in more detail on pp. 78-79. It seems clear, however, that the problem of discontinuous distribution on serpentine may best be explained in terms of the interplay of changing climate on the genetics of populations (evolution) during spans of geologic time. Stebbins (1942a) has pointed out that most rare and narrowly endemic species in continental areas are genetically homogeneous, consisting of relatively few biotypes adapted to a few sites that usually are restricted geographically. By contrast, widespread species frequently are genetically heterogeneous with many biotypes grouped into more or less distinct ecotypes of wide distribution. As he notes, these relations may be explained in two ways (Stebbins, 1942a). First, the restricted species was once abundant, widespread and rich in biotypes, and its present rarity is due to the depletion of its store of genetic variability: these are "depleted species," or paleoendemics. The taxa noted above in Groups I and II provide examples of how species may gradually become depleted and confined to more local areas. The second possible origin for rare taxa is that the species was never common, but originated from a small population of a widespread taxon following establishment in a local area: these are "insular species," either on an island, or an outcrop of serpentine, or other unique rock in a given area. Representative of this mode are relatively new taxa in rapidly proliferating genera, as in Group III.

There are various grades of paleoendemics. Some have no close relatives, as *Adenostoma, Carpenteria,* or *Lyonothamnus.* Others have their nearest relatives in distant regions and may be generically distinct, as *Taxodium-Sequoia-Metasequoia-Cryptomeria* of the Taxodiaceae. In addition, paleoendemics may be represented by ecotypes that live in environments quite different from those of their nearest fossil relatives, as the Miocene and Pliocene ecotypes allied to *Lyonothamnus asplenifolius* Greene, *Populus tremuloides* or *Picea breweriana* (Axelrod, 1940, 1976b). The fossil record shows that many present-day paleoendemics were common and widespread into the Pliocene. They gradually became rare in California as biotypes and ecotypes were eliminated as summer rains decreased, the depletion process resulting in their confinement to a few favorable, more local environments.

On this basis, some of the plants that are now found chiefly on ultrabasic rocks probably had a wide distribution during the Miocene and Pliocene, like the taxa listed above in Group I. They evidently were restricted subsequently to serpentine areas by their inability to compete with taxa on non-serpentine sites as climate gradually changed.

Discussion. – We have now see that most serpentine areas were exposed after the Miocene ($\sim$ 10 m.y.), and that many are later Pliocene or younger; that numerous taxa that have considerable antiquity (15-16 m.y.) occur today on serpentine and many other substrates, and ranged widely over non-serpentine areas in the past; and finally, that all taxa that are now in the region adapted gradually to decreasing summer rainfall during Miocene and later times. With these points in mind, we can now consider an hypothesis that appears to explain the discontinuous occurrence of taxa on serpentine.

Experimental results reported by Walker (1954) and Kruckeberg (1954), and reviewed by Proctor and Woodell (1975), show that populations on serpentine may be distinct ecotypically from those of the same taxa on non-serpentine sites. They not only have a tolerance of low calcium levels, but also low levels of K^+, PO_4^{--}, and other essential nutrients, and they can also tolerate Co^{++}, Cr^{+++}, Ni^{++} and other toxic ions. Manifestly, if adaptation to serpentine is followed by biotype and ecotype depletion in non-serpentine areas, the result will be isolated populations restricted to serpentine. This could be accomplished if the taxa were distributed widely on non-serpentine sites in the Miocene and Pliocene, and were eliminated there as summer rains decreased. On this basis, the populations on ultrabasic rocks would be relict. They persist because of the poorer competitive ability there of taxa in adjacent vegetation on non-serpentine substrates that have a normal representation of mineral nutrients for their growth and reproduction.

In regard to the survival of relict taxa on serpentine, moisture may also be a factor in their persistence. Contrary to a prevalent notion, serpentine areas are not dry: they are quite moist to wet. This is apparent from (1) the streams that flow longer in serpentine areas than in adjacent terrains; from (2) the occurrence of moist meadows in these areas; from (3) numerous slides and seepages on highways constructed through serpentine areas; from (4) the regular and often abundant occurrence of *Umbellularia* on serpentine outcrops, as well as scattered clumps of *Mimulus* and other annual mesophytes that are still flowering at small seepages even late in the year; from (5) the common deposition of silica (opal:$SiO_2 \cdot nH_2O$) on serpentine clasts that lie on the surface which reflect ample capillary water rising to the surface, and also from (6) the abundance of clay in serpentine outcrops which are the result of intense chemical weathering owing to ample water in the soil. The physiological significance (if any) of high moisture content for ecotypes of woody endemics that are confined chiefly to serpentine outcrops is not clear. However, it may compensate for the lack of summer rain and together with nutrient conditions, enable them to escape competition from the native flora that gradually adapted to emerging mediterranean climate during the later Pliocene and Early Pleistocene.

In summary, some of the species like *Cupressus sargentii* that are widely discontinuous on ultrabasic rocks in California appear to be paleoendemics. It is proposed that more widely distributed ecotypes of these species disappeared as summer rains decreased during the Pliocene and Early Pleistocene, a phenomenon that also restricted ecotypes of widespread Pliocene species that have survived in California. The present discontinuous areas of *Cupressus sargentii, Quercus durata* and other taxa are therefore but remnants of earlier, more widespread distributions on "ordinary" as well as on ultrabasic rocks. It is significant that *C. sargentii* is most nearly related to *C. goveniana* Gord. and *C. abramsiana* C. B. Wolf, narrow endemics of the Monterey Peninsula and the central Santa Cruz Mountains, respectively (Griffin and Critchfield, 1972). Since both taxa are non-serpentine in their occurrence, the relation is consistent with the suggestion that *C. sargentii* probably was widely spread on non-serpentine substrates earlier in its history. In view of this relationship, the frequent occurrence of "northern," and often mesophytic species on serpentine, as noted by Hardham (1962) for groves of *Cupressus sargentii* in the Santa Lucia Mountains, is expectable. Such examples afford a possible explanation for the relict occurrence of the narrow endemic *Cercocarpus traskiae*, known now from only a few small trees in Salta Verde Canyon on the south coastal plain of Catalina Island at an elevation of about 250 m (Thorne, 1969, p. 63). Related to evergreen species now confined to central Mexico, it is restricted to a very basic outcrop of saussuritized olivine-clinopyroxene gabbro at its only occurrence on the island (Platt, 1976). *Cercocarpus traskiae* is clearly a relict of the Madrean woodland vegetation which had invaded California by the Oligocene, and has persisted here where it is removed from competition with the rest of the flora (Axelrod, 1967b, p. 277).

Some of these discontinuous distributions may have resulted from the disruption of a formerly continuous woodland cover, whereas others might be related to distributions attained in the locally moist, open conditions of the Pleistocene. The disjunct distributions of taxa such as those in *Arctostaphylos, Ceanothus* sect. *Cerastes, Hesperolinon,* and the *Streptanthus glandulosus* complex may have developed in a similar way, but in these groups they have been modified by their own special genetic features, as discussed in the next section.[2]

MODES OF EVOLUTION

Numerous books and review papers discuss the general outlines of plant evolution, and here we shall focus on some of the major patterns that are characteristic of California, and especially of the California Floristic Province. In view of the analysis presented above, they seem to be directly responsible for much of the proliferation of species, including endemics, in this region. First, we shall review the evolution of species in homogamic complexes, especially of woody plants, and then the evolution of annual species, mostly dicots.

Woody Plants and Perennial Herbs

Hybrids between species of woody plants and perennial herbs are often fertile (Raven, 1976b). This seems to have made possible the relatively rapid reassortment of the genetic material of differentiated species in such groups and hence, in large measure, their readjustment in response to changing conditions against the background of a fluctuating environment (Epling, 1947b; Anderson, 1949, 1953; Anderson and Stebbins, 1954; Stebbins, 1969). They constitute what Grant (e.g., 1971) has termed homogamic complexes—hybrid complexes in which the derived species have originated by processes of hybrid speciation that do not involve drastic changes in the genetic system. Such processes seem to have been very important in California, and there are numerous examples of the proliferation of species by this means (Lewis, 1969b, 1972). We shall begin with a discussion of *Ceanothus* sect. *Cerastes,* one of the best studied.

As pointed out by Johnston (1971), *Ceanothus* is one of the least specialized genera of Rhamnaceae, and has no close relatives. There are two sections, *Ceanothus* and *Cerastes.* Sect. *Ceanothus* is widespread in Mexico and Central America and reaches the eastern United States and southern Canada, with 32 species, 24 in California, and 21 endemic to the California Floristic Province. Sect. *Cerastes,* very distinct in its bizarre sunken stomatal crypts and many species with opposite leaves, has 21 species, 20 in California, and 17 endemic to the California Floristic Province. All of them, so far as known, lack the ability to crown-sprout, although some may be relatively long-lived (Keeley, 1975). Two of the species in California range northward to Oregon and Washington, and one, *C. greggii* A. Gray, occurs south to Oaxaca and east to Texas. The only species in sect. *Cerastes* not found in California, *C. lanuginosus* (Jones) Rose, occurs in the mountains from Chihuahua to San Luis Potosí.

Although intersectional hybrids are very difficult to obtain, Nobs (1963) demonstrated that all species of sect. *Cerastes* are interfertile. He studied in detail 11 species occurring in the North Coast Ranges of California, and found that they inhabit a great diversity of substrates. Two were restricted to serpentine, with *C. jepsonii* having a highly discontinu-

---

2. A recent valuable contribution to our understanding of edaphic endemism is that of H. Wild and A. D. Bradshaw, *Evolution* 31: 282-293 (1977).

ous distribution on it throughout the area; five were confined to Pliocene Sonoma volcanics of varied composition; and one restricted to Franciscan marine sandstones of Cretaceous age. Within this complex environmental mosaic, the ability of species of sect. *Cerastes* to hybridize and produce hybrid recombinants better suited to particular combinations of ecological factors than their parents as the climate changed has doubtless given the group a great adaptive advantage. Such varied ecologic situations became more frequent in the North Coast Ranges from the Pliocene onward as tectonism increased. The ability of hybrids to exploit disturbed areas was pointed out by Anderson (1948). In turn, this has led to a multiplication of species in this restricted area of a typically Madrean group that was already in existence in Middle Miocene time (17 m.y. BP).

A similar pattern has been described for *Mimulus* sect. *Diplacus* by Beeks (1962), in which five species in southern California form a complex much like that described by Nobs for *Ceanothus* sect. *Cerastes*. Observations on species of *Salvia* sect. *Audibertia* in the same region suggest a similar role for hybridization in the evolution of that group (Epling, 1938, 1947a, 1947b; Anderson and Anderson, 1954; Epling et al., 1962; Webb and Carlquist, 1964; Grant and Grant, 1964). In the same way, hybridization has probably played a role in the evolution of most genera listed in table 15, and especially, as pointed out by Wells (1968, 1969), among the non-crown-sprouting species of *Arctostaphylos* (e.g., Epling, 1947b; Gottlieb, 1968; Keeley, 1976; Roof, 1976), a genus in which all 50 species occur in the California Floristic Province, although one has attained a circumboreal distribution and is disjunct in Guatemala. That hybridization between ecotypes may be an important factor in the evolution of a single species is suggested by the studies of *Adenostoma fasciculatum* H. & A. by Edgar Anderson (1954). On the other hand, Young (1974), while reporting introgression in *Rhus*, did not regard it as important in the evolution of that group. Patterns suggestive of reticulate evolution at the diploid level also occur in the following additional woody genera of California in each of which only a single chromosome number is known, among others: *Abies* (Hamrick and Libby, 1972; Hamrick, 1976), *Amelanchier, Berberis (Mahonia), Cornus, Cupressus* (Wolf, 1948; Lawrence et al., 1975), *Eriodictyon, Fraxinus* (Miller, 1955), *Holodiscus, Keckiella, Lavatera, Lepechinia, Malacothamnus* (Bates, 1963), *Opuntia* subg. *Platyopuntia* (Benson and Walkington, 1965), *Pinus* (Stockwell and Righter, 1946; Stebbins, 1950, p. 209-210; Zobel, 1951; Libby, 1958; Haller, 1961, 1962; Critchfield, 1966, 1967; Smith, 1971; Lanner, 1974), including subsect. *Oocarpae, Quercus* (Tucker, 1952, 1953, Benson et al., 1967), *Ribes, Sambucus,* and *Solanum umbelliferum* sens. lat. (W. F. Hinton, personal communication). In several additional groups, the corresponding patterns probably also occur, but are accompanied by polyploidy: for example, the *Chaenactis douglasii* (Hook.) H. & A. complex (Mooring, 1965), *Epilobium (Zauschneria) canum* (Greene) Raven (Raven, 1976a), *Eriophyllum* (Mooring, 1973, 1975), *Lupinus* (e.g., Nowacki and Dunn, 1964), and *Salix.*

Wells (1969) has pointed out that the ability to sprout from the base following a fire is undoubtedly the primitive condition in the evergreen sclerophylls that make up the chaparral vegetation of California and Baja California (table 15). This ability has been lost by some taxa of *Arctostaphylos* and *Ceanothus*, the most speciose of these genera. Wells has argued convincingly that the loss of the ability to crown-sprout in about four-fifths of the taxa of these genera has been associated with a more rapid adaptation to local conditions, and hence the development of numerous species. In the crown-sprouting taxa, on the other hand, evolution has been more conservative and ecotypic differentiation has tended to

TABLE 15

Relation between Mode of Reproduction in Response to Fire and Extent of Differentiation of
Taxa among the Evergreen Sclerophylls in the 21 Woody Genera Making up the
Chaparral Vegetation of California and Baja California (Modified from Wells, 1969).

Genus	Crown-Sprouting Taxa	Nonsprouting Taxa
Arctostaphylos	16 (21.3%)	59 (78.7%)
Ceanothus	12 (20.7%)	46 (79.3%)
Quercus	12	0
Rhamnus	7	0
Fremontodendron	6	0
Garrya	6	0
Lonicera	5	0
Cercocarpus	4	0
Adenostoma	3	0
Dendromecon	3	0
Chrysolepis	2	0
Comarostaphylis	2	0
Heteromeles	2	0
Pickeringia	2	0
Prunus subg. *Laurocerasus*	2	0
Rhus subg. *Schmaltzia*	2	0
Cneoridium	1	0
Malosma	1	0
Ornithostaphylos	1	0
Simmondsia	1	0
Xylococcus	1	0

be the rule. For example, only eight of the 50 species of *Arctostaphylos* crown-sprout
(P. V. Wells, personal communication), and these are the most polymorphic in the genus.

The origin of new species at the diploid level following hybridization is by no means
confined to the woody members of the California flora. A convincing case has been made
for the origin of *Delphinium gypsophilum* Ewan of the inner South Coast Ranges (Lewis
and Epling, 1954, 1959) in this way. Ornduff (1969b) describes the interesting case of the
annual *Lasthenia burkei* (Greene) Greene, which might be a hybrid derivative of two other
diploid ($n = 6$) species with which it is freely crossable. Grant has discussed in detail the
important role of hybridization in the evolution of the annual leafy-stemmed gilias (Grant,
1953). Straw (1955, 1956) has outlined the origin of a hybrid species of *Penstemon* in
southern California, in which differential pollination appears to have played a significant
role. In *Helianthus*, Heiser (1949) has traced the origin of weedy races of *H. bolanderi* A.
Gray through introgression from *H. annuus* L.; non-introgressed populations of *H. bolanderi* are strictly confined to serpentine soils. Gottlieb (1971, 1972) has discussed the annual species of *Stephanomeria* as a homogamic complex in which polyploidy has also
played a role; *Camissonia* sect. *Chylismia* (Raven, 1962), in contrast, is a homogamic complex with very little polyploidy. Tillett (1967) has documented the occurrence of widespread hybridization and introgression in the maritime species of *Abronia*. All species of
Iris sect. *Californicae*, with a gametic chromosome number $n = 20$, are interfertile, and the
elegant analysis of Lenz (1958) makes it seem highly likely that hybridization has played
a role in the origin of some of the entities recognized in nature at present. The same may

be said of a number of other plant groups, among them *Calochortus, Calystegia, Cardamine* sect. *Dentaria* (insufficient chromosomal information), *Castilleja* (accompanied by polyploidy), *Cirsium* (polyploidy), *Corethrogyne*, the *Dodecatheon hendersonii* A. Gray complex (Thompson, 1953), *Dudleya* (polyploidy), *Erysimum, Heuchera, Lilium, Lotus,* and *Monardella* (polyploidy), to name just a few examples. It has been scarcely 20 years since the hypothesis of the origin of diploid entities following hybridization between species first began to be discussed seriously, and no doubt many additional examples will be found as biosystematic studies are pursued.

As to polyploidy itself as an evolutionary force in California, little can be added to the perceptive discussions of Stebbins and Major (1965) and Stebbins (1971, p. 179-190). Stebbins (1971, p. 185) has calculated the percentage of polyploidy for north coastal California at about 36%, neither strikingly high nor strikingly low. As it becomes possible to assess percentages of polyploidy for various habitats within the State, significant relationships may emerge, but these are certainly not apparent at present. A number of excellent studies of polyploid complexes have been made on Californian plants, notably those of Clausen, Keck, and Hiesey on *Penstemon* (1940) and Madiinae (1945), that of Stebbins (1942b) on the *Eriogonum fasciculatum* Benth. complex, that of Heckard (1968) on *Castilleja*, those of Lewis and his associates on *Clarkia*, an annual genus (e.g., Lewis, 1953a, 1953b; Lewis and Lewis, 1955; Abdel Hameed, 1971), and those of Dempster and her co-workers on *Galium* (e.g., Dempster and Ehrendorfer, 1965; Dempster and Stebbins, 1968).

Little also can be said about apomixis, which appears to play a very minor role in the flora of California, where it occurs in genera such as *Antennaria, Arabis, Crepis* (Babcock and Stebbins, 1938), *Poa,* and *Potentilla* (Clausen, Keck and Hiesey, 1940). In these genera, no large swarms of species comparable with those in *Crataegus* and *Rubus* in temperate eastern North America occur. The native Californian species of the latter two genera are uniformly sexually reproducing.

Among the examples of biosystematic investigations of perennial groups that have contributed to our knowledge of the California flora are those on *Achillea* (Clausen, Keck and Hiesey, 1948), *Brodiaea* (Niehaus, 1971), *Calyptridium* (Hinton, 1975, 1976a, 1976b), *Dichanthelium* (Spellenberg, 1975), *Elymus* (Snyder, 1950, 1951), *Grindelia* (Dunford, 1964), *Hulsea* (Wilken, 1975), *Jepsonia* (Ornduff, 1969c), *Malacothamnus* (Bates, 1963), *Mimulus* sect. *Erythranthe* (Hiesey, Nobs and Bjorkman, 1971), *Oenothera* (Klein, 1970), *Penstemon* (Clausen, Keck and Hiesey, 1940), *Potentilla* (Clausen, Keck and Hiesey, 1940; Clausen and Hiesey, 1958), *Scrophularia* (Shaw, 1962), *Sidalcea* (Kruckeberg, 1957), *Silene* (Kruckeberg, 1955, 1961), the blue-flowered species of *Sisyrinchium* (Henderson, 1976), and *Epilobium* sect. *Zauschneria* (Clausen, Keck and Hiesey, 1940). Although relatively few such groups have been studied in biosystematic detail, they have begun to illuminate the pattern of evolution among such plants in California.

Annual Herbs

Critical biosystematic studies of many genera of plants in California have led to an increased understanding of them, and of the relative importance of phenomena such as autogamy, allopolyploidy, and saltational speciation in them. The high number of species of annual plants in California is directly associated with these phenomena. One of the first groups of plants to be studied in biosystematic detail was the subtribe Madiinae of Asteraceae (Clausen, Keck and Hiesey, 1949; Clausen, 1951). Other genera of annual California plants that have received comparable attention include *Blennosperma* (Ornduff, 1964), *Calycadenia* (Carr, 1975), *Collinsia* (e.g., Garber, 1960; Garber and Gorsic, 1956; Garber and Bell, 1962; Ahloowalia and Garber, 1961; Bell and Garber, 1961; Garber and Dhillon, 1962), *Downingia* (Wood, 1961), *Gayophytum* (Thien, 1969), *Gilia* (summary in Grant, 1971), *Lasthenia* (Ornduff, 1966, 1969b, 1976; Altosaar, Bohm, and Ornduff, 1974), Limnanthaceae (Ornduff, 1969a, 1971; Ornduff and Crovello, 1968; Arroyo, 1973; Jain, 1976), *Lupinus* (Dunn, 1956; Harmon and Dunn, 1968), *Mentzelia* (Thompson, 1960 and unpublished), *Microseris* (Chambers, 1955), *Mimulus* (e.g., Vickery, 1959, 1967, 1969, 1974; Tai and Vickery, 1970; Alam and Vickery, 1973), *Myosurus* (Stone, 1959), *Nemophila* (Cruden, 1971), *Orthocarpus* (Atsatt, 1970), *Phacelia* (Gillett, 1955), and *Stephanomeria* (Gottlieb, 1971, 1972). In many of these groups of annuals, experimental hybridization showed that hybrid sterility occurred both within and between taxa; a good example in *Clarkia* is provided by the studies of Mosquin (1964). The "biological species concept" is therefore of limited value as a means of describing the pattern of variation in these groups (Raven, 1976b). Exceptional is *Lessingia*, in which barriers to hybridization are scarce or absent and hybrids are fertile, both within and between species (Spence, 1963). The origin of some species of annual plants in California has been explained by hybridization between diploid species, as mentioned on pp. 80-81.

In some groups of annuals, of which *Plectritis* (Dempster, 1958; Morey, 1959) and *Vulpia* (Lonard and Gould, 1974) are good examples, autogamy has led to the perpetuation of certain character combinations that have been given undue taxonomic weight until the studies cited led to a more meaningful reevaluation. In other cases, however, such as *Mentzelia* sect. *Trachyphytum* (Thompson and Roberts, 1974 and unpublished), *Clarkia* sect. *Myxocarpa* (Mosquin, 1962; Small, 1971a, 1971b), *Camissonia* (Raven, 1969), the *Claytonia perfoliata* Donn complex (Miller, 1976), and *Gilia* (summary in Grant, 1971), biosystematic studies have led to the recognition of many previously undetected autogamous, mostly polyploid entities and to an increase in the number of recognized species.

The notion of saltational speciation, the result of catastrophic selection, was developed by Lewis (1962, 1966, 1969b, 1973) as a result of his studies of evolution in the Californian genus *Clarkia*. In a variable and fluctuating climate, occasional extreme reduction of population size may be associated with reorganization of the gene pool. Such drastic fluctuations are characteristic of the populations of annual plants in California (cf. Talbot, Biswell, and Hormay, 1939). This may give rise to a new population which is at once reproductively and spatially isolated from the parental population and endowed with characteristics whereby it may be uniquely successful in a new habitat. Since marginal populations often grow in edaphic situations unusual for the species as a whole, saltational speciation frequently gives rise to rare, derivative species of narrow edaphic requirements (Raven, 1964). Furthermore, saltational speciation may be most important in the annual species of mediterranean climates, in which total precipitation may fluctuate greatly, in which seed longevity is usually limited, and in which the genetic constitution of the population may therefore change readily if it is reduced to a very low number—as in periods of drought (Bartholomew, Eaton and Raven, 1973). This will be especially true if they are colonial. In perennial species, such as *Delphinium*, or desert annuals, such as *Linanthus parryae* (A. Gray) Greene (Epling and Dobzhansky, 1942; Epling et al., 1960), and *Camissonia* sect. *Chylismia*, dormant genotypes may express themselves in subsequent years to "swamp" the effects of temporary changes owing to particular unfavorable or unusual years (Raven, 1962). Gottlieb (1973a, 1974b) has stressed the fact that in those plants in which seeds last in the soil and may germinate over a period of years, the kinds of fluctuation in population size that seem to have led to saltational speciation in *Clarkia* may not be operative. In *Clarkia*, all seeds germinate following autumn rains and survival is critical; if the subsequent rains are inadequate, the population has a high probability of becoming extinct. Clearly, the limits of the phenomenon of saltational speciation need to be explored further.

The derivation of *Clarkia franciscana* Lewis & Raven from *C. rubicunda* (Lindl.) Lewis & Lewis was one of the first instances to be explained by saltational speciation (Lewis and Raven, 1958), although the term itself was not coined until several years later. Recently, Gottlieb (1973b) has shown that *C. franciscana* is not similar to *C. rubicunda* in the alleles it has fixed at six out of the eight enzyme systems that he studied, and that it is probably not derived from *C. rubicunda* in the direct fashion as envisioned by Lewis and Raven (1958). On the other hand, the derivation of *C. lingulata* Lewis & Lewis ($n = 9$) from *C. biloba* (Dur.) Nels. & Macbr. subsp. *australis* Lewis & Lewis ($n = 8$), as outlined by Lewis and Roberts (1956), is strongly supported by similar electrophoretic data (Gottlieb, 1974a). Other examples seem to include the origins of *Clarkia exilis* Lewis & Vasek (Vasek, 1958), *C. springvillensis* Vasek, *C. tembloriensis* Vasek, and *C. "Caliente"* from *C. unguiculata* Lindl. (Vasek, 1971; Lewis, 1973) and the origin of a distinctive, marginal, self-pollinating population of *C. xantiana* A. Gray studied by Moore and Lewis (1965). All these evolutionary events in *Clarkia*, and the whole pattern of evolution in the genus, appear to be correlated with progressive adaptation to increasing aridity (Lewis, 1953a, 1953b; Vasek, 1964; Vasek and Sauer, 1971; Davis, 1970), accompanied by aneuploid changes in chromosome number, allopolyploidy, and autogamy.

The role of saltational speciation in genera other than *Clarkia* is difficult to assess, and the phenomenon itself is difficult to document (Lewis, 1972). It has been invoked by Kyhos (1965) to explain the origin of two annual species of *Chaenactis* (Asteraceae) on the deserts from a third species that occurs in more mesic sites in cismontane California. In the self-incompatible *Calycadenia pauciflora* A. Gray, saltational changes involving several more or less simultaneous chromosomal rearrangements seem to have accompanied the origin of several races, although most of the chromosomal and morphological differences within the complex seem to have accumulated gradually (Carr, 1975). *Hesperolinon*, a highly self-pollinating complex of annual species in which population size fluctuates drastically (Sharsmith, 1961, p. 249), certainly exhibits a pattern of variation suggestive of saltational speciation. Grant (1971) has also interpreted some of the patterns he observed in *Gilia* as resulting from processes of this sort. The role of saltational speciation in the evolution of woody and perennial plant groups is yet to be determined (Lewis, 1966).

INTRODUCED PLANTS

The introduction of exotic plants into California began in May, 1769, when Father Junipero Serra reached San Diego Bay and founded the first permanent European settlement in Alta California. Hendry (1931) has even inferred that *Rumex crispus* L., *Erodium cicutarium* (L.) L'Hér., and *Sonchus asper* (L.) Hill were introduced earlier, but we agree with Frenkel (1970) that this seems unlikely. At any event, Frenkel (1970) has calculated that at least 16 species of exotic plants were established in California during Spanish colonization, 1769-1824; 63 additional species during Mexican occupation, 1825-1848; and 55 more during American pioneer settlement, 1849-1860. Thus there were at least 134 alien species established in California by 1860. Parish (1920) listed 186 immigrant species, excluding fugitives and waifs, for southern California alone, and Jepson (1925) considered that there were at the time he wrote 292 species of established exotics that constituted an integral part of the native flora. Robbins et al. (1951) listed 437 introduced species of weeds in California, but treated many of them rather superficially.

The exotic flora of California at the present time is difficult to assess for a number of reasons. Despite this, the importance of naturalized plants in the vegetation is obvious; at lower elevations they often make up 50 or 75% of the flora (Jepson, 1925). Munz and Keck (1959) listed 797 species of introduced plants, and Munz later added 178 more (Munz, 1968), for a total of 975 species (Howell, 1972). Concerning this impressive figure, we agree with Howell (1972) that the representation of species of introduced plants has been greatly inflated by the inclusion of several categories of plants that do not constitute genuine elements in the flora of the State. Many new weeds are being established in California with each passing decade, and some (e.g., *Iris pseudacorus* L., Raven and Thomas, 1970) are spreading aggressively; but there are others which do not appear properly to constitute a part of the flora of the State.

Owing to the equable climate of coastal California and the regular, widespread use of water during the summer dry period, many species do persist in the immediate vicinity of gardens and near cultivated individuals, or even become established from seed locally (see discussion in Howell et al., 1958). They should be collected and recorded when found, but they should not be considered members of the flora of the State until they have become fully established away from cultivation, or until it is clear that they are capable of persisting without human interference. Among the genera that we consider to fall into this category are the following: *Albizia, Althaea, Anchusa, Aptenia, Arctotis, Axonopus, Borago, Cedronella, Cestrum, Chasmanthe, Conicosia, Consolida, Coprosma, Cosmos, Crocosmia, Cymbalaria, Dianthus, Disphyma, Drosanthemum, Duchesnea, Eucalyptus, Eugenia, Freesia, Gazania, Glaucium, Gunnera, Haloragis, Hebe, Hedera, Hesperis, Hunnemannia, Hyparrhenia, Ionopsidium, Lantana, Leonotis, Libertia, Lunaria, Nicandra, Origanum, Parkinsonia, Phlomis, Romulea, Ruta, Santolina, Schinus, Scorzonera, Stenotaphrum, Ulmus, Venidium,* and *Vinca,* as well as the Australian species of *Acacia* and such well known species of garden plants as *Antirrhinum majus* L., *Asclepias curassavica* L., *Buddleja davidii* Franchet, *Centaurea cineraria* L., *Cyperus alternifolius* L., *Dimorphotheca eklonis* DC., *Geranium robertianum* L., *Heliotropium amplexicaule* Vahl, *Hypericum canariense* L., *Mirabilis jalapa* L., *Oenothera speciosa* Nutt., *Opuntia tomentosa* Salm-Dyck, *Papaver bracteatum* Lindl., *Petunia violacea* Lindl., *Salix babylonica* L., *Salvia grahamii* Benth., *Saxifraga sarmentosa* L., *Senecio cruentus* DC., *Sorbus aucuparia* L., *Tagetes patula* L., *Verbena tenuisecta* Briq., *Viola odorata* L., and *V. tricolor* L. Although some of these plants, and others like them in California, may be fully established at present, we have not found evidence that this is the case. The same is true of all species of *Pelargonium* (nine) except *P. grossularioides* (L.) Ait., all seven exotic species of *Prunus* listed by Munz and Keck (1959), a number of species of *Ipomoea* and many other entities accorded full status in his *Flora.* We are especially grateful to T. C. Fuller for his comments on these lists, with which he concurs.

The distinction between such species and those accorded full status is by no means always clear, but we agree with Howell (1972) that a conservative interpretation is preferable when treating the flora of the State. It is difficult to specify the status of species such as *Oxalis hirta* L., *O. martiana* Zucc., *O. purpurea* Thunb., and *O. rubra* St. Hil., always found in or near gardens. Such species as *Echium fastuosum* Ait. and *Tropaeolum majus* L., to name but two examples, have spread from gardens to become well established in nature. On the other hand, there are a number of species well established and frequent in California which are inevitably found in the vicinity of gardens and towns, such as *Amaranthus deflexus* L., *Chenopodium multifidum* L., *Coronopus didymus* (L.) Sm., *Cotula australis* (Sieb.) Hook.f., *Euphorbia peplus* L., *Lobularia maritima* (L.) Desf., *Polycarpon tetraphyllum* (L.) L., and even *Stellaria media* (L.) Vill.

Some weedy species in the flora of California are usually or always found in lawns, where they are watered fairly regularly through the summer; even though several of these are common, one wonders whether they would persist without such watering. Examples are *Agrostis tenuis* Sibth., *Carex leavenworthii* Dewey, *Cyperus brevifolius* (Rottb.) Hassk., *Hydrocotyle sibthorpioides* Lam., and *Stellaria graminea* L. Other species, frequent in lawns, also occur away from them in meadowy habitats in the mountains or near the coast, such as *Bellis perennis* L., *Cerastium vulgatum* L., and *Sherardia arvensis* L.

Leaving aside the special problems of weeds that occur mainly in orchards [e.g., *Emex spinosa* (L.) Campd., *Tagetes minuta* L.], cultivated fields (e.g., *Orobanche ramosa* L.), irrigated pastures and alfalfa fields [*Sphaerophysa salsula* (Pall.) DC.], or grainfields [e.g., *Galium tricornutum* Dandy (*G. tricorne* Stokes pro parte), *Myosotis virginica* (L.) BSP.], there are a number of crop plants that persist for limited periods of time, and which also have been recorded as members of the flora of California, including *Anethum, Armoracia, Asparagus, Beta, Cannabis, Citrullus, Coriandrum, Fagopyrum, Humulus, Lens, Lycopersicum, Oryza, Pastinaca, Petroselinum, Pisum,* and *Triticum,* and such species as *Avena sativa* L., *Cichorium endivia* L., *Cynara scolymus* L., *Daucus carota* L., *Glycyrrhiza glabra* L., *Hordeum vulgare* L., *Linum usitatissimum* L., *Sorghum vulgare* Pers., *Vicia faba* L., and *Vitis vinifera* L. We believe that all of these should be excluded from the flora of the State, but note at the same time that a few other crop plants, such as *Brassica oleracea* L., *Medicago sativa* L., *Raphanus sativus* L., and *Secale cereale* L., have become well established.

There are a number of introduced species regarded as members of the flora of California which have not been collected for decades and are not likely to have persisted in the State. Examples are *Ambrosia trifida* L., *Briza media* L., *Camelina microcarpa* Andrz., *Centaurea eriophora* L., *Coronopus squamatus* (Forsk.) Asch., *Dyssodia papposa* (Vent.) Hitchc., *Echinops sphaerocephalus* L., *Emex australis* Steinh., *Erysimum cheiranthoides* L., *Lycopsis arvensis* L., *Malva verticillata* L., *Mantisalca salmantica* (L.) Briq. & Cavillier (*Centaurea salmantica* L.), *Melampodium perfoliatum* HBK., *Oxybaphus nyctagineus* (Michx.) Sweet, *Potentilla recta* L., *Rapistrum rugosum* (L.) All., *Sieglingia decumbens* (L.) Bernh., *Silene dichotoma* Ehrh., and *Solanum sisymbriifolium* Lam. The great majority of these species are no more likely to be encountered in California than hundreds of others which have never been recorded in the State, and we believe that to include them in standard floras merely makes these books less useful for their primary purposes than they would be otherwise.

Species known only from one or two localities present a special problem in evaluation. Plants such as *Amaranthus spinosus* L., *Atriplex vesicaria* Heward, *Brassica fruticulosa* Cyr.,

Brayulinea densa (Willd.) Small, *Carex molesta* Mkze., *Cyanopsis muricata* (L.) Dostál (*Centaurea muricata* L.), *Flaveria trinervia* (Spreng.) C. Mohr, *Geranium anemonifolium* L'Hér., *Rumex kerneri* Rorb., *R. stenophyllus* Ledeb., *Trifolium agrarium* L., and *Vitex agnus-castus* L. may either decline, persist at their single localities, or spread and achieve floristic prominence; meanwhile, decisions about the desirability of including them as standard members of the flora of the State must of necessity be subjective, and based in part upon a subjective evaluation of their potential for increase.

Analysis

With these factors in mind, we have scrutinized the list of introduced plants presented by Munz and Keck (1959) and by Munz (1968) in the light of various local floras such as Howell (1970) and Howell et al. (1958) and our own field knowledge, and have calculated that the introduced flora of the state of California consists of two genera and two species of ferns, approximately 249 genera and 504 species of dicots, and about 87 genera and 168 species of monocots, of which 65 genera and 137 species are grasses: a total of 338 genera and 674 species. This suggests that about 301 species, or 31% of the total recorded by Munz and Keck (1959) and Munz (1968), should not be regarded as members of the flora of California. The monocots comprise about 24.9% of the total, considerably higher than for the native flora of the State (18.1%) entirely because of the rich representation of grasses, which constitute 20.3% of the total, as weeds. In the area of Gray's Manual (Fernald, 1950), monocots comprise 15% of the total introduced plants, 28.2% of the native plants: dicots are more successful as weeds.

Only about 6 genera and 11 species of introduced plants in California occur only outside of the boundaries of the California Floristic Province, so the two totals can be taken as identical, for all practical purposes. Both genera of ferns, 136 of the 249 genera of dicots, and 50 of the 87 genera of monocots are not found in California as native plants. The families Dipsacaceae, Eriocaulaceae, Loganiaceae, Tamaricaceae, and Tropaeolaceae are known in California only as introduced plants, and Resedaceae would be added to this list if *Oligomeris* were interpreted as introduced in North America. The families Aponogetonaceae, Basellaceae, Cannabaceae, Commelinaceae, Meliaceae, Myrtaceae, and Phytolaccaceae, as well as Asteraceae, tribe Arctotideae, listed for California by Munz and Keck (1959) and by Munz (1968), are here excluded from the naturalized flora of the State on the basis of the reasoning presented above.

Of the 504 species of naturalized dicots, two are trees (*Ailanthus, Robinia*), five are vines (*Lonicera,* 2 species; 1 each of *Lathyrus, Parthenocissus, Solanum*), 29 are shrubs (18 genera), 172 are biennials and perennials, and 296 are annuals—58.7% annuals. Of the 168 species of naturalized monocots, 80 are perennials and 88 are annuals—52.4% annuals. For the naturalized flora of California as a whole, 57% of the species are annuals, a very impressive figure and one that should be compared with the weedy floras of other parts of the world. The representation of annuals in the weedy flora of California is similar to the figures mentioned above for the floras of the drier parts of the Near East and North America.

Of the 674 species of introduced plants in California, there are 559 of Old World origin and only 115 of New World origin. Among the New World species, some 53 are found in the United States, 25 in tropical America, and 38 in South America. The Old World species include 18 from Australasia, 30 from central and southern Africa, at least 19 from tropical Asia, and the remainder—492 species—from temperate to arid Eurasia and North Africa. Thus, some 73%, or nearly three-quarters, of the naturalized flora of California is from Eurasia and North Africa, with very little of that from eastern Asia; 4% from central and southern Africa; about 2.5% each from Australasia and tropical Asia; and only 16.8% from the New World. With the comprehensive modern floras, such as *Flora Europaea* and *Flora USSR*, that are now available, it should be possible to analyze the native ranges of the introduced plants of California with much greater precision in the future, and to compare the weedy flora of California on this basis with similar floras in other parts of the world.

A number of groups of introduced plants in California are very "difficult" taxonomically. An improved understanding of them can probably come only from more detailed taxonomic studies within the areas where they are native. Notable among these are the annual species of *Bromus, Centaurea, Chenopodium, Polygonum, Salsola,* and the groups of species centering around *Galium aparine* L., *Rumex acetosella* L., and *Solanum nigrum* L., the last group including both native and introduced entities. It is very much to be hoped that California botanists will utilize the taxonomy and nomenclature for their naturalized weeds that have been worked out in such standard floras as *Flora Europaea*, since it is rarely possible to gain greater insight into a group of plants where it is naturalized than where it is native.

A few species in California appear to include both native and introduced strains: *Achillea millefolium* L. (if interpreted broadly), *Calystegia sepium* (L.) R. Br., *Mentha arvensis* L., *Poa compressa* L., *P. pratensis* L., *Prunella vulgaris* L. (Nelson, 1965), and *Veronica serpyllifolia* L. Many other species have been regarded as native, only subsequently to be found to have been introduced; examples are two species of *Mesembryanthemum* (Moran, 1950a); *Cakile* (Rodman, 1974); and *Vulpia myuros* (L.) K. C. Gmelin var. *hirsuta* Hack., a plant of European origin earlier regarded as a native species and generally called *Festuca megalura* Nutt. (Lonard and Gould, 1974). *Oligomeris* and *Plantago ovata* have been discussed on p. 42. Hybridization between introduced and native species has produced complex patterns of variation in *Amaranthus* (Tucker and Sauer, 1958) and in *Opuntia* (Benson and Walkington, 1965), and perhaps in other groups as well. In some cases, as in *Barbarea*, the introduced and native species are so similar as to cause serious confusion in identification.

It is often difficult to deal with the taxonomy of introduced plants because the information available from other parts of the world may be insufficient. Thus *Hordeum glaucum* Steud. was recognized as a distinct species in California and named *H. stebbinsii* Covas (Covas, 1949) before it was realized that an earlier name was available. *Polygonum marinense* Mertens & Raven, described as an endemic of Marin County, California (Mertens and Raven, 1965), might occur in the Old World and be introduced in California, but it is not clear whether another name is available for it or not. The "pappose form" of *Centaurea pouzinii* DC. mentioned by Powell et al. (1974) is in fact a polyploid which, although very likely of European origin, may or may not have a name at present. Finally, introduced plants sometimes behave differently in the lands they are colonizing. *Erodium brachycarpum* (Godr.) Thell. [*E. obtusiplicatum* (Maire, Weiller & Wilcz.) J. T. Howell (Baker, 1962; Robertson, 1972; Major, 1975)], abundant in California, is occasional in North Africa and very rare in Mediterranean Euope; it was apparently introduced to California from Chile, and to Chile from Spain. *Bromus mollis* (Knowles, 1943; Jain,

Marshall, and Wu, 1970) is an autogamous species that may have differentiated genetically since it was first introduced into California and neighboring regions. Differentiation
has apparently also occurred within 50 years in the relatively uncommon populations
of *Lythrum tribracteatum* Salzm. ex Ten., a strongly inbreeding species introduced from
the Mediterranean region (Baker and Baker, 1976). In *Raphanus,* hybridization between
two introduced species in California has modified the pattern of variation in both
(Panetsos and Baker, 1968).

Among the native species of California are some which have become weedy during the
past century. Robbins et al. (1951) list 256 native weeds, and Stebbins (1965) has provided a useful discussion of 41 which have distinct colonizing tendencies. Some, like *Amsinckia, Claytonia perfoliata, Eschscholzia californica* Cham., *Lupinus arboreus* Sims, and
Mimulus guttatus DC., have become weeds on other continents. These plants provide interesting opportunities for experimental studies. It is clear, however, that several millenia
of cultivation around the Mediterranean, in the Near East, and in North Africa, possibly
together with the higher proportion of annual species in the arid parts of the Old World,
have resulted in the evolution of the vast majority of the weeds now in California.

In summary, the naturalized flora of California consists of about 674 species, of which
168 (24.9%) are monocots, 137 (20.3%) are grasses, and 112 (16.6%) are Asteraceae. As
compared with the flora of the State as a whole, grasses and Asteraceae are both over-rep-
resented, with grasses, which constitute only 309 species or about 6% of the flora of the
State, very much so. In this connection, it is interesting to note that 60% of the naturalized
grasses in California, but only 24.6% of the native grasses, are annuals. Only two grasses,
one annual, are listed among the 41 most successful native weeds of California by Stebbins (1965). Otherwise, the naturalized weed flora of California, probably like other weed
floras in temperate regions at least, has a high proportion of dicots and annual dicots especially—a relationship that recalls the statistics for the native flora of California as a
whole.

Because of the special problems concerned with the taxonomic treatment of weedy
plants and their economic importance, the preparation of a weed flora of California would
be highly desirable. Standard floras often state the ranges of weeds in vague terms, perhaps
because the weeds are expected to attain wider ranges in due course, and collectors, if
they do preserve specimens of weeds, often neglect to record the kind of detail that would
be most helpful in evaluating their status. A new and critical inventory of the weeds of
California would provide an important stimulus to the study of the flora of the State. The
considerable advantages of weedy plants for many kinds of evolutionary studies have been
stressed by Baker (1974), among others. Some outstanding modern studies have been con-
ducted on the weeds of California, such as those of Iman and Allard (1965), Jain and Marshall (1967), Jain (1969), Hamrick and Allard (1972, 1975), and others on *Avena,* and
others reviewed by Baker (1974). In addition, comprehensive investigations like that of
Frenkel (1970) on the roadside vegetation of California are badly needed. Considering
their great economic importance, our ignorance of most aspects of the biology of weeds
is truly profound.

PRESERVATION OF THE CALIFORNIA FLORA

No account of the plants of California would be complete without mentioning the increasing threat to many of the species caused by urbanization, agriculture, and pollution. Many species of *Calochortus, Dudleya, Fritillaria,* and *Lilium* are in imminent danger of extinction by those who take them from the wild to gardens, in which they may persist for only a short time. Certain cacti, as well as *Agave utahensis* Engelm. var. *nevadensis* Engelm. ex Greenm. & Roush, and *Nolina interrata* Gentry, are actually exploited commercially. An informed public should bring these practices to a halt at once. Native plants should be introduced into cultivation only by means of seeds and cuttings, and the rare plants of California should be stringently protected in the wild.

Of the endemic native plants of California, some 28 species and 5 additional infraspecific taxa are apparently extinct already in the wild (Powell, 1974; Ripley, 1975; Ayensu, 1975). *Cordylanthus palmatus* (Ferris) Macbr. has been preserved in cultivation and is possibly reestablished now at a habitat near those where it once grew (Chuang and Heckard, 1973). In addition to these, 156 species and 80 additional infraspecific taxa in California are considered endangered, and 279 species and 71 additional subspecies are seriously threatened (Ripley, 1975; Ayensu, 1975). The national list must continue to be reviewed in the light of local data such as that presented by Powell (1974), and will doubtless continue to be modified in the light of new information as well as the changing status of the plants. At any rate, some 459 species and 157 additional infraspecific taxa in California, about one out of every ten species in the total flora, are either already extinct or are in danger of extinction. Looking at these figures in another way, about a quarter of all the threatened and endangered plants in the United States occur in California. This is not surprising in view of the extreme specialization and narrow restriction of many taxa in California. More and more taxonomic papers, like those of Wilson (1970, 1972), Chuang and Heckard (1973), and Jain (1976), will no doubt include discussions of the probabilities of extinction of some or all the taxa they are treating. It would certainly seem that the citizens of California would want to preserve their irreplaceable heritage of native plants, but botanists must take the responsibility of informing them about this critical problem.

DIRECTIONS FOR THE FUTURE

Some of the most interesting studies of California plants are those directed at the population level. Although they may not directly affect the taxonomy of the species concerned, they will in time provide valuable insight into the evolution of those groups, and of plants in general. Some of these studies have already been mentioned in our discussions of serpentine endemism and of saltational speciation. Into this category also might be placed the studies of the Carnegie Institution group, which have done so much to clarify our understanding of the origin of ecotypes (e.g., Clausen, Keck, and Hiesey, 1940, 1948; Clausen and Hiesey, 1958; Hiesey, Nobs, and Björkman, 1971). The flora of California, and particularly the annual dicots, will continue to furnish excellent material for a variety of population genetic and evolutionary studies in the future, as they have in the past. Examples of such studies are those on *Linanthus* (Epling and Dobzhansky, 1942; Epling, Lewis, and Ball, 1960), *Lupinus* (e.g., Harding and Mankinen, 1967, 1971), *Ambrosia* (Payne et al., 1973), *Vulpia* (Kannenberg and Allard, 1967), *Stephanomeria* (Gottlieb, 1973b, 1975), *Collinsia* (Weil and Allard, 1965), and of several genera by Linhart (1975). Such investigations are also close to some aspects of physiological ecology, an extremely promising field which is still in its infancy.

Further application of the methods of numerical taxonomy to the plants of California will make possible a new precision in understanding their relationships. Among the first studies of this sort to be carried out on California plants are those of Crovello (1968) on *Salix*; Ornduff and Crovello (1968), Arroyo (1973) and Parker (1977) on Limnanthaceae; McNeill (1975) on Portulacaceae, tribe Montieae; Bartholomew et al. (1973) on *Clarkia*; Nelson and van Horn (1976) on *Pentachaeta*; and Wilken (1977) on *Hulsea*.

In understanding the patterns of evolution in the California flora, we are fortunate in having as complete a set of biosystematic investigations as are available for any flora. Nevertheless, many more comparative studies should be carried out if comprehensive information is to be developed. It would be extremely useful if detailed biosystematic studies, involving both intra- and interspecific hybridization, could be carried out on more groups of woody plants, comparable to those of Nobs (1963) on *Ceanothus* sect. *Cerastes*. These studies should be carried out in botanical gardens, such as those at Berkeley, Claremont, Stanford, or Santa Barbara, since they take a considerable amount of space and time for their completion. Nevertheless, they would clarify much about the evolution of the flora of California and the nature of the groups involved. Among the herbaceous groups that appear to us to be particularly suitable for such investigations are: *Acanthomintha, Calochortus, Calystegia, Chorizanthe, Collomia, Corethrogyne,* the "*Dentaria californica* Nutt." complex, *Dudleya, Eriastrum, Erysimum, Eschscholzia* (annual species), *Githopsis, Hesperolinon, Lewisia, Linanthus, Monardella, Navarretia, Nemacladus, Orcuttia, Parvisedum, Pectocarya, Plagiobothrys, Pogogyne, Polygonum* sect. *Duravia,* and *Sidalcea* (annual species). A few groups have special interest not only because of the biosystematic relationships within California but also because of their relationships with species found in distant areas. Outstanding in this category are *Antirrhinum, Calandrinia, Fagonia, Helianthemum, Lavatera, Menodora, Styrax,* and the annual genera of Asteraceae, tribe Inuleae. In these cases, attention should be given to patterns of variation in the populations and to the factors important in evolution, with the clarification of the taxonomy a secondary goal.

There are also many genera, well developed in California but widespread elsewhere, which should be studied to clarify their evolutionary patterns. This notably concerns the way in which spreading aridity from Upper Tertiary times in the western United States, culminating in the development of a full-mediterranean climate in the later Quaternary, has affected their evolution. Genera in this category include: *Agropyron, Agrostis, Allium, Arabis, Arenaria, Arnica, Asclepias, Aster, Astragalus, Atriplex, Bromus, Calamagrostis, Carex, Castilleja, Chamaesyce, Chrysopsis, Cirsium, Cryptantha, Elymus, Epilobium, Erigeron, Eriogonum, Eryngium, Erythronium, Fritillaria, Gnaphalium, Heuchera, Hosackia, Ivesia, Juncus, Lilium, Lomatium, Lotus (Hosackia), Lupinus, Mimulus* (except the better studied sects. *Erythranthe* and *Simiolus*), *Muhlenbergia, Phacelia, Potentilla, Salix, Sanicula, Senecio, Stachys, Stipa,* and *Zigadenus.* Although some biosystematic studies of a few of these groups have been carried out, none is really well known in this sense in California.

In this regard, the question arises as to whether it is possible to relate the evolution of certain genera to the history of the San Andreas fault, which has had an average rate of movement of about 2.5 cm/yr during the past 12 million years. If systematic differences are exhibited by populations on opposite sides of the rift, if one or more of them have a fossil record, and if there is reason to believe that the taxa have limited powers of migration, then it might be possible to estimate rate of evolution in the group, providing that migration in response to climatic change has not further complicated the picture. As an example, in *Pinus* subsect. *Oocarpae* populations east and west of the rift are specifically distinct. Geologic evidence implies that the taxa now in coastal California-northern Baja California and in the Sierra Madre Occidental of Mexico were in proximity into the Late Miocene. That the taxa probably are older, and have not undergone significant change since then, is consistent with the fossil record as now known (Axelrod, 1967a).

For further advances in understanding, it would be most desirable if a computerized bank of information about the characteristics and distributions of the plants of California could be established, possibly under State Government auspices. Only in this manner can the complex matrix of facts concerning about 5,720 native and introduced species of vascular plants be marshalled effectively. The questions that might be asked of such a data bank are not only systematic and evolutionary ones, but increasingly practical ones that may provide valuable insight into ecological and even social phenomena of great interest and commercial importance. Patient revisions of the existing floras will no longer suffice for the needs of science and society.

Finally, it is apparent that to place our systematic and evolutionary studies in proper time perspective, further studies of Neogene and Quaternary floras are especially needed. Large and still unstudied collections are in the museums at Berkeley, Stanford, and no doubt elsewhere that can add materially to our understanding of the history of the flora and of the vegetation of California.

SUMMARY

The native flora of California consists of 154 families, 878 genera, and 5046 species of vascular plants, of which 26 genera (3.0%) and 1517 species (30.1%) are endemic. For the California Floristic Province, defined as that part of the State west of the Sierra Nevada-Cascade axis and excluding the deserts, together with adjacent portions of southwestern Oregon and the nondesertic northwestern corner of Baja California, there are 795 genera, 50 (6.3%) endemic, and 4452 species, 2125 (47.7%) endemic. The proportion of dicots, and especially of annual dicots, is unusually high; and it is the dicot groups in which a proliferation of species has taken place in recent geologic time that have made the greatest contribution to the high number of species and high proportion of endemism in the flora of California. In general, the flora of California resembles the weedy flora of the world in its high proportion of annuals and, like it, has evolved in response to recent climatic change and periodic drought.

In a historical context, the flora of California can be thought of as comprising two elements: a northern, or Arcto-Tertiary one, and a southern, or Madro-Tertiary one, a relationship first perceived by Asa Gray a century ago. Within the broad ecotone between these fundamentally different vegetation types, a summer-dry climate developed following the Pliocene and provided a major stimulus for the proliferation of species and some genera. More than half of the genera and species in the California Floristic Province have Arcto-Tertiary affinities. Another quarter of the genera and a third of the species are of Madro-Tertiary origin. An additional 20% of the genera and 15% of the species are also of

Madrean origin, but from its drier regions, and having entered the California Floristic Province mainly during the Quaternary. The remaining 5% of the genera, with a small number of species, have come from miscellaneous sources.

The relict genera and species occur in the areas both where Arcto-Tertiary and where Madro-Tertiary elements predominate but are best represented in areas of relatively high summer rainfall. Their survival in California has been conditioned by the equability of the climate of the coastal portions of the region which provided a haven for evolutionary oddities that were often more widespread in the past. In contrast, the neoendemics are best represented in areas of full mediterranean climate, especially those in ecotonal situations, where they have developed rapidly, mostly in response to the vicissitudes of an oscillating wet-dry climate during the past several million years.

Six families, 121 genera, and 938 species occur within the borders of the state of California but not in the California Floristic Province. The deserts in which they occur are recent in their present regional extent, but analogous pockets of semiarid vegetation that probably occurred farther south included many of their ancestors. The Inyo region, with at least seven endemic genera, stands out floristically among these desert regions, with many endemic species, especially on the steep limestone cliffs near Death Valley. Plants growing in the crevices and shade of such cliffs receive greater effective moisture than those on the rough level ground nearby, and may in at least some cases be relicts that have survived locally from times when the climate was moister.

Among the groups that are especially prominent in the California Floristic Province and significant in terms of their endemism are Amaryllidaceae, tribe Allieae, subtribe Brodiaeinae, with 39 of the 45 total species in the California Floristic Province and 32 of them endemic; Hydrophyllaceae, with 99 species, including 65 endemics in the California Floristic Province, of a total of 250 species; Onagraceae, with 135 of 650 species in the California Floristic Province; Polemoniaceae, with 163 of 317 species in California and 23% of the entire family, 74 species, endemic in the California Floristic Province; and Polygonaceae, subfamily Eriogonoideae, with 12 of the 14 genera and 159 of the 321 species in California, 80 endemic in the California Floristic Province. Of these six groups, Boraginaceae may basically be northern, but all of the others appear to be Madrean and either autochthonous (Brodiaeinae) or ultimately of tropical and/or South American origin (Hydrophyllaceae, two tribes of Onagraceae, Eriogonoideae, two tribes of Polemoniaceae). None appears to have originated within or to have radiated primarily within the California Floristic Province, Brodiaeinae being the most likely to have had such an origin; this underscores the recency of summer-dry and other extreme climates of the region in which the hundreds of species enumerated above have evolved. Indeed, the California Floristic Province may be likened to an island (of new climates) which has been colonized by plants from other areas that have sometimes radiated extensively there and contributed to the unique floristic character of the region.

The upper elevations of the Sierra Nevada and the arid flats of the Central Valley are geologically recent habitats with relatively few endemics and almost no relicts, whereas the middle elevations of the Central Coast Ranges, the Islands, and northwestern Baja California are ecotonal regions of equable climate that are inhabited by many endemics, both ancient and recently derived. The Klamath-Siskiyou Region, with its varied topography and relatively abundant warm-season precipitation, and the North Coast region in general, are outstanding centers in which relict species have survived.

Edaphic endemism has been important in the derivation of the flora of California, with serpentine and associated ultrabasic rocks playing a key role. Many of the species that are wholly or partly restricted to such substrates are older than the development of the soils in which they occur, and appear to have been progressively restricted to these soils on account of their high moisture-holding capacity as the climate has become progressively drier. In a number of actively evolving groups, however, neoendemics also originated on these soils during the Quaternary.

Among the woody plants and perennial herbs of California, the origin of new species following interspecific hybridization in homogamic complexes has been frequent, as exemplified by *Ceanothus* sect. *Cerastes*. In this respect, the California Floristic Province again resembles an island (of newly formed climate) in which there have been abundant recent opportunities for evolution. Annual dicots, in contrast, have produced scores of new species, especially in ecotonal areas, through combinations of autogamy, polyploidy, and saltational speciation, and have contributed most numerically to the present constitution of the California flora.

In the Late Pliocene, before the final elevation of the Sierra Nevada-Cascade axis, the Transverse Ranges and the Peninsular Ranges, the area of the present California Floristic Province was probably inhabited by about 25% fewer species than at present, perhaps about 3,400. Only about a third of these, rather than half, would have been endemic. These figures are based upon those characteristic of other regions of similar size that lie in temperate latitudes and are not characterized by a mediterranean climate. Primarily owing to the multiplication of species of annual dicots, but secondarily to their proliferation through hybridization in woody and perennial genera, the flora has attained its present level of some 4,452 species. Either a trend toward increasing moisture or one toward increasing aridity would cause many of these neoendemics to be lost, and further deterioration of the climate would eliminate many of the relicts also.

The naturalized flora of California consists of some 338 genera and 674 species, 57% of which are annuals. In contrast, the native flora of the State includes only 28.6% annuals. Some 300 additional species are sometimes regarded as members of the naturalized flora of the State but should be eliminated because they are not known to reproduce freely away from cultivation, occur only as waifs, or have not been collected for many years. About three-quarters of the naturalized flora of the State is from Eurasia and North Africa, some 17% from the New World, and the remainder from miscellaneous sources; relatively few of the native plants of California have been successful as weeds in California or elsewhere.

In view of the wealth of biosystematic knowledge available about the plants of California, they should provide ideal subjects for study in the developing field of plant population biology, which is concerned mainly with evolution at the population level. Many groups are still in need of biosystematic attention, and there is much to be learned in this region of rapid evolution against a gradient of high environmental relief and fluctuating climate. Computerization of the available information about the plants of California would be highly desirable and would provide one key to further conceptual and practical advance. Further detailed studies of Neogene and Quaternary plant fossils, many of which have already been collected and await study, will continue to teach us much about the history of the richest, most diverse, and most highly endemic flora of the temperate regions of the Western Hemisphere.

ACKNOWLEDGMENTS

We are grateful to H. K. Airy Shaw, D. M. Bates, R. Y. Berg, S. T. Blake, D. Boufford, L. Constance, W. B. Critchfield, T. C. Croat, D. B. Dunn, C. Durrell, L. A. Galloway, A. Gentry, W. F. Grant, A. V. Hall, L. R. Heckard, W. F. Hinton, R. H. Holm, J. Holub, M. P. Johnson, M. C. Johnston, R. M. King, A. R. Kruckeberg, R. M. Lloyd, II. E. Moore, Jr., M. A. Nobs, J. L. Reveal, R. G. Stolze, D. Taylor, J. M. Tucker, D. H. Valentine, D. C. Wasshausen and P. V. Wells for information they kindly provided us during the course of this study; and to T. C. Fuller, A. R. Kruckeberg, J. T. Howell, R. Ornduff and G. L. Stebbins for their useful reviews of the manuscript. Reid Moran has been especially generous in sharing with us a wealth of unpublished data concerning the plants of Baja California, and has also provided many useful comments on the manuscript.

Our studies have been supported by a series of independent grants from the National Science Foundation.

LITERATURE CITED

ABDEL-HAMEED, F.
  1971.   Cytogenetic studies in *Clarkia*, section *Primigenia*. V. Interspecific hybridization between *C. amoena huntiana* and *C. lassenensis*. Evolution 25: 347-355.

ABRAMS, L.
  1925.   The origin and geographical affinities of the flora of California. Ecology 6: 1-6.
  1926.   Endemism and its significance in the California flora. Proc. Internat. Congr. Plant Sci., Ithaca 2: 1520-1523.

ADAMSON, R. S. and T. M. SALTER.
  1950.   Flora of the Cape Peninsula. Juta, Cape Town and Johannesburg. xix + 889 pp.

AHLOOWALLA, B. S. and E. D. GARBER.
  1961.   The genus *Collinsia*. XII. Cytogenetic studies of interspecific hybrids involving species with pediceled flowers. Bot. Gaz. 122: 219-228.

ALAM, M. T. and R. K. VICKERY, JR.
  1973.   Crossing relationships in the *Mimulus glabratus* heteroploid complex. Amer. Midl. Nat. 90: 449-454.

AL-SHEHBAZ, I. A.
  1973.   The biosystematics of the genus *Thelypodium* (Cruciferae). Contr. Gray Herb. Harvard Univ. 204: 1-148.

ALT, K. S. and V. GRANT.
  1960.   Cytotaxonomic observations on the goldback fern. Brittonia 12: 153-170.

ALTOSAAR, I., B. A. BOHM, and R. ORNDUFF.
  1974.   Disc-electrophoresis of albumin and globulin fractions from dormant achenes of *Lasthenia*. Biochem. Syst. Ecol. 2: 67-72.

ANDERSON, E.
  1948.   Hybridization of the habitat. Evolution 2: 1-9.
  1949.   Introgressive Hybridization. Wiley & Sons, New York. 109 pp.
  1953.   Introgressive hybridization. Biol. Revs. Cambridge 28: 280-307.
  1954.   Introgression in *Adenostoma*. Ann. Missouri Bot. Gard. 41: 339-350.

ANDERSON, E. and B. R. ANDERSON.
  1954.   Introgression of *Salvia apiana* and *S. mellifera*. Ann. Missouri Bot. Gard. 41: 329-338.

ANDERSON, E. and G. L. STEBBINS.
  1954.   Hybridization as an evolutionary stimulus. Evolution 8: 378-388.

ANONYMOUS.
  1941.   Climate and Man. Yearbook of Agriculture. U.S. Dept. of Agriculture. 1248 pp.
  1957-68. Geologic Map of California. Scale 1:250,000. Calif. Div. Mines and Geol. 27 sheets.

ARROYO, M. T. K.
  1973.   A taximetric study of infraspecific variation in autogamous *Limnanthes floccosa* (Limnanthaceae). Brittonia 25: 177-191.

ATSATT, P. R.
  1970.   The population biology of annual grassland hemiparasites. II. Reproductive patterns in *Orthocarpus*. Evolution 24: 598-612.

AXELROD, D. I.
  1940.   The concept of ecospecies in Tertiary paleobotany. Proc. Natl. Acad. Sci. U.S. 27: 545-551.
  1944.   The Sonoma flora (California). Carnegie Inst. Wash. Publ. 553: 167-206.
  1950.   The evolution of desert vegetation in western North America. Carnegie Inst. Wash. Publ. 590: 215-306.
  1956.   Mio-Pliocene floras from west-central Nevada. Univ. Calif. Publ. Geol. Sci. 33: 1-316.
  1958.   Evolution of the Madro-Tertiary Geoflora. Bot. Rev. 24: 433-509.
  1966a. The Eocene Copper Basin Flora of northeastern Nevada. Univ. Calif. Publ. Geol. Sci. 59: 1-83, pl. 1-20.
  1966b. The Pleistocene Soboba flora of southern California. Univ. Calif. Publ. Geol. 60: 1-79, pl. 1-14.

1967a. Evolution of the Californian closed-cone pine forest. *In* R. N. Philbrick (ed.), Proceedings of the Symposium on the Biology of the California Islands, pp. 93-149. Santa Barbara Botanic Garden, Santa Barbara, California.

1967b. Geologic history of the Californian insular flora. *In* R. N. Philbrick (ed.), Proceedings of the Symposium on the Biology of the California Islands, pp. 267-316. Santa Barbara Botanic Garden, Santa Barbara, California.

1967c. Drought, diastrophism, and quantum evolution. Evolution 21: 201-209.

1968.   Tertiary floras and topographic history of the Snake River basin, Idaho. Geol. Soc. Amer. Bull. 79: 713-734.

1971.   Fossil plants from the San Francisco Bay region. *In* Geologic Guide to the Northern Coast Ranges, Point Reyes Region, California. Geol. Assoc. Sacramento Guidebook, pp. 74-86.

1973.   History of the Mediterranean ecosystem in California. *In* F. di Castri and H. Mooney (eds.), Mediterranean Type Ecosystems, Origin and Structure, pp. 225-277. Springer-Verlag, New York-Heidelberg-Berlin.

1975.   Evolution and biogeography of Madrean-Tethyan sclerophyll vegetation. Ann. Missouri Bot. Garden 62: 289-334.

1976a. Evolution of the Santa Lucia fir (*Abies bracteata*) ecosystem. Ann. Missouri Bot. Garden 63: 24-41.

1976b. History of the coniferous forests, California and Nevada. Univ. Calif. Publ. Botany 70: 1-62.

1977.   Outline history of California vegetation. *In* M. Barbour and J. Major (eds.), Terrestrial Vegetation of California, pp. 139-220. John Wiley Interscience, New York.

AYENSU, E. S.
1975.   Manuscript changes in Ripley, 1975.

BABCOCK, E. G. and G. L. STEBBINS, JR.
1938.   The American species of *Crepis*. Their interrelationships and distribution as affected by polyploidy and apomixis. Carnegie Inst. Wash. Publ. 504: 1-199.

BAILEY, V. L.
1962.   Revision of the genus *Ptelea* (Rutaceae). Brittonia 14: 1-45.

BAKER, H. G.
1953.   Dimorphism and monomorphism in the Plumbaginaceae. III. Correlation of geographical distribution patterns with dimorphism and monomorphism in *Limonium*. Ann. Bot. n.s. 17: 615-627.

1962.   Weeds–native and introduced. Jour. Calif. Hort. Soc. 23: 97-104.

1974.   The evolution of weeds. Ann. Rev. Ecol. Syst. 5: 1-24.

BAKER, I. and H. G. BAKER.
1976.   Variation in an introduced *Lythrum* species in California vernal pools. Univ. Calif. Davis Inst. Ecol. Publ. 9: 63-69.

BARKLEY, F. A.
1937.   A monographic study of *Rhus* and its immediate allies in North and Central America, including the West Indies. Ann. Missouri Bot. Gard. 24: 265-498.

BARNEBY, R. C.
1964.   Atlas of North American *Astragalus*. 2 vols. New York Bot. Garden.

BARNEBY, R. C. and E. C. TWISSELMANN.
1970.   Notes on *Loeflingia* (Caryophyllaceae). Madroño 20: 398-408.

BARTHOLOMEW, B., L. C. EATON, and P. H. RAVEN.
1973.   *Clarkia rubicunda*: a model of plant evolution in semiarid regions. Evolution 27: 505-517.

BASSETT, I. J. and B. R. BAUM.
1969.   Conspecificity of *Plantago fastigiata* of North America with *P. ovata* of the Old World. Canad. Jour. Bot. 47: 1865-1868.

BATES, D. M.
1963.   The genus *Malacothamnus*. Ph.D. Dissertation, Univ. Calif., Los Angeles; University Microfilms, Inc., Ann Arbor, Mich. 171 pp.

1968.   Generic relationships in the Malvaceae, tribe Malveae. Gentes Herbarum 10: 117-135.

BATES, D. M. and O. J. BLANCHARD, JR.
1970.   Chromosome numbers in the Malvales. II. New or otherwise noteworthy counts relevant to classification in the Malvaceae, tribe Malveae. Amer. Jour. Bot. 57: 927-934.

BEATLEY, J. C.
  1976. Vascular Plants of the Nevada Test Site and Central-Southern Nevada: Ecologic and Geographic Distributions. Energy Research and Development Administration. viii + 308 pp.
BEEKS, R. M.
  1962. Variation and hybridization in southern California populations of *Diplacus* (Scrophulariaceae). Aliso 5: 83-122.
BELL, C. R.
  1954. The *Sanicula crassicaulis* complex (Umbelliferae). A study of variation and polyploidy. Univ. Calif. Publ. Bot. 27: 133-230, pl. 9-13.
BELL, S. L. and E. D. GARBER.
  1961. The genus *Collinsia*. XII. Cytogenetic studies of interspecific hybrids involving species with sessile flowers. Bot. Gaz. 122: 210-218.
BEMIS, W. P. and T. W. WHITAKER.
  1969. The xerophytic *Cucurbita* of northwestern Mexico and southwestern United States. Madroño 20: 33-41.
BENSON, L., E. A. PHILLIPS, P. A. WILDER, et al.
  1969. The Native Cacti of California. Stanford Univ. Press, Stanford, Calif. xii + 243 pp.
BENSON, L., E. A. PHILLIPS, P. A. WILDER, ET AL.
  1967. Evolutionary sorting of characters in a hybrid swarm. I: Direction of slope. Amer. Jour. Bot. 54: 1017-1026.
BENSON, L. and D. L. WALKINGTON.
  1965. The southern California prickly pears—invasion, adulteration, and trial-by-fire. Ann. Missouri Bot. Gard. 52: 262-273.
BRECKON, G. J. and M. G. BARBOUR.
  1974. Review of North American Pacific Coast beach vegetation. Madroño 22: 333-360.
BRANDEGEE, K.
  1893. Sierra Nevada plants in the Coast Range. Zoe 4: 168-176.
BROOME, C. R.
  1974. Systematics of *Centaurium* (Gentianaceae) of Mexico and Central America. Ph.D. Dissertation, Duke Univ., Durham, N.C. xiii + 415 pp. University Microfilms, Ann Arbor, Mich. 74-25,390.
  1977. The Central American species of *Centaurium* (Gentianaceae). Brittonia 28: 413-426.
BRUMMITT, R. K.
  1974. A remarkable new species of *Calystegia* (Convolvulaceae) from California. Kew Bull. 29: 499-502.
CAMPBELL, D. H. and I. L. WIGGINS.
  1947. Origins of the flora of California. Stanford Univ. Publ. Biol. Sci. 10: 1-20.
CARLQUIST, S.
  1959. Studies on Madinae: Anatomy, cytology, and evolutionary relationships. Aliso 4: 171-236.
CARR, G. D.
  1975. *Calycadenia hooveri* (Asteraceae), a new tarweed from California. Brittonia 27: 136-141.
  1976. Chromosome evolution and aneuploid reduction in *Calycadenia pauciflora* (Asteraceae). Evolution 29: 681-699.
CHABOT, B. F. and W. D. BILLINGS.
  1972. Origins and ecology of the Sierran alpine flora and vegetation. Ecol. Monogr. 42: 163-199.
CHAMBERS, K. L.
  1955. A biosystematic study of the annual species of *Microseris*. Contr. Dudley Herb. 7: 207-312.
CHANEY, R. W.
  1954. A new pine from the Cretaceous of Minnesota and its paleoecological significance. Ecology 35: 145-151.
CHANEY, R. W. and H. L. MASON.
  1930. A Pleistocene flora from Santa Cruz Island, California. Carnegie Inst. Wash. Pub. 415: 1-24, pl. 1-7.
CHASE, V. C. and P. H. RAVEN.
  1975. Evolutionary and ecological relationships between *Aquilegia formosa* and *A. pubescens* (Ranunculaceae), two perennial plants. Evolution 29: 474-486.

CHRTEK, J. and J. HOLUB.
  1963.   Poznámky k taxonemii a nomenklatuře rodů *Evax* Gaertn. a *Filago* L. Preslia 35: 1-17.
CHUANG, T.-I. and L. CONSTANCE.
  1969.   A systematic study of *Perideridia* (Umbelliferae-Apioideae). Univ. Calif. Publ. Bot. 55: 1-74.
CHUANG, T.-I. and L. R. HECKARD.
  1973.   Taxonomy of *Cordylanthus* subgenus *Hemistegia* (Scrophulariaceae). Brittonia 25: 135-158.
CLAPHAM, A. R., T. G. TUTIN, and E. F. WARBURG.
  1962.   Flora of the British Isles. Ed. 2. Cambridge Univ. Press, London. xlviii + 1269 pp.
CLAUSEN, J.
  1951.   Stages in the Evolution of Plant Species. Cornell Univ. Press, Ithaca, N.Y. viii + 206 pp.
CLAUSEN, J. and W. M. HIESEY.
  1958.   Experimental studies on the nature of species. IV. Genetic structure of ecological races.
          Carnegie Inst. Wash. Publ. 615: i-iv, 1-312.
CLAUSEN, J., D. D. KECK, and W. M. HIESEY.
  1940.   Experimental studies on the nature of species. I. Effect of varied environments on western
          North American plants. Carnegie Inst. Wash. Publ. 520: i-vii, 1-452.
  1945.   Experimental studies on the nature of species. II. Plant evolution through amphiploidy and
          autoploidy, with examples from the Madiinae. Carnegie Inst. Wash. Publ. 546: i-vii, 1-174.
  1948.   Experimental studies on the nature of species. III. Environmental responses of climatic races
          of *Achillea*. Carnegie Inst. Wash. Publ. 581: i-iii, 1-129.
CLAUSEN, R. T.
  1975.   *Sedum* of North America North of the Mexican Plateau. Cornell Univ. Press, Ithaca and
          London. 742 pp.
CLEMENTS, F. E.
  1920.   Plant indicators. The relation of plant communities to process and practice. Carnegie Inst.
          Wash. Publ. 290: i-xvi, 1-388.
  1936.   The origin of the desert climax and climate. *In* T. H. Goodspeed (ed.), Essays in Geobotany
          in Honor of William Albert Stechell, pp. 87-140. Univ. California Press, Berkeley.
CLOKEY, I. W.
  1951.   Flora of the Charleston Mountains, Clark County, Nevada. Univ. Calif. Publ. Botany 24:
          1-274.
CONSTANCE, L.
  1952.   *Howellianthus*, a new subgenus of *Phacelia*. Madroño 11: 198-203.
  1963.   Amphitropical relationships in the herbaceous flora of the Pacific Coast of North and South
          America: A symposium. Introduction and historical review. Quart. Rev. Biol. 38: 109-116.
CORE, E. L.
  1941.   The North America species of *Paronychia*. Amer. Midl. Nat. 25: 369-397.
CORRELL, D. S. and M. C. JOHNSTON.
  1970.   Manual of the Vascular Plants of Texas. Texas Research Foundation, Renner. xv + 1881 pp.
COVAS, G.
  1949.   Taxonomic observations on the North American species of *Hordeum*. Madroño 10: 1-21.
CRAMPTON, B.
  1954.   Morphological and ecological considerations in the classification of *Navarretia* (Polemonia-
          ceae). Madroño 12: 225-238.
  1954.   Observations on the genus *Soliva* in California. Leafl. West. Bot. 7: 196-198.
  1959.   The grass genera *Orcuttia* and *Neostapfia*: A study in habitat and morphological specializa-
          tion. Madroño 15: 97-110.
  1961.   The endemic grasses of the California Floral Province. Leafl. West. Bot. 9: 154-158.
CRISTÓBAL, C. L.
  1960.   Revision del género "*Ayenia*" L. Op. Lilloana 4: 1-230.
CRITCHFIELD, W. B.
  1966.   Crossability and relationships of the California big-cone pines. U.S. For. Serv. Res. Pap. NC-6:
          36-44.
  1967.   Crossability and relationships of the closed-cone pines. Silvae Genet. 16: 89-97.
CROAT, T. C.
  1977.   Flora of Barro Colorado Island, Panama. Stanford Univ. Press, Stanford, Calif. (in press).

CRONQUIST, A.
1950. A review of the genus *Psilocarphus*. Res. Stud. St. Coll. Washington 18: 71-89.
CRONQUIST, A., A. H. HOLMGREN, N. H. HOLMGREN, and J. L. REVEAL.
1972. Intermountain Flora, Vol. 1. Hafner, New York. iii + 370 pp.
CROVELLO, T. J.
1968. A numerical taxonomic study of the genus *Salix*, section *Sitchenses*. Univ. Calif. Publ. Bot. 44: 1-61.
CRUDEN, R. W.
1971. Information on chemistry and pollination biology relevant to the systematics of *Nemophila menziesii* (Hydrophyllaceae). Madroño 21: 505-515.
DAMMER, U.
1889. Polygonaceae. Natürl. Pflanzenfam. III. 1a: 1-36.
DANIN, A.
1972. Mediterranean elements on rocks of the Negev and Sinai deserts. Notes Roy. Bot. Gard. Edinburgh 31: 437-440.
DANIN, A., G. ORSHAN, and M. ZOHARY.
1975. The vegetation of the northern Negev and the Judean Desert of Israel. Israel Jour. Bot. 24: 118-172.
DAUBENMIRE, R.
1975. Floristic plant geography of eastern Washington and northern Idaho. Jour. Biogeogr. 2: 1-18.
DAVIDSON, C.
1973. An anatomical and morphological study of Datiscaceae. Aliso 8: 49-110.
DAVIS, W. S.
1970. The systematics of *Clarkia bottae, C. cylindrica*, and a related new species, *C. rostrata*. Brittonia 22: 27-284.
DAVIS, W. S. and H. J. THOMPSON.
1967. A revision of *Petalonyx* (Loasaceae) with a consideration of affinities in subfamily Gronovioideae. Madroño 19: 1-18.
DEMPSTER, L. T.
1958. Dimorphism in the fruits of *Plectritis*, and its taxonomic implications. Brittonia 10: 14-27.
1973. The polygamous species of the genus *Galium* (Rubiaceae), section *Lophogalium*, of Mexico and southwestern United States. Univ. Calif. Publ. Bot. 64: 1-36.
1975. An Asian *Kelloggia* (Rubiaceae). Madroño 23: 100-101.
DEMPSTER, L. T. and F. EHRENDORFER.
1965. Evolution of the *Galium multiflorum* complex in western North America. II. Critical taxonomic revision. Brittonia 17: 289-334.
DEMPSTER, L. T. and G. L. STEBBINS.
1968. A cytotaxonomic revision of the fleshy-fruited *Galium* species of the Californias and southern Oregon (Rubiaceae). Univ. Calif. Publ. Bot. 46: 1-52, 2 pls.
DETLING, L. E.
1961. The chaparral formation of southwestern Oregon, with considerations of its postglacial history. Ecology 42: 348-357.
1968. Historical background of the flora of the Pacific Northwest. Bull. Mus. Nat. Hist. Univ. Oregon 13: 1-57.
DUNFORD, M. P.
1964. A cytogenetic analysis of certain polyploids in *Grindelia* (Compositae). Amer. Jour. Bot. 51: 49-60.
DUNN, D. B.
1956. The breeding systems of *Lupinus*, group *Micranthi*. Amer. Midl. Nat. 55: 443-472.
1971. A case of long range dispersal and "rapid speciation" in *Lupinus*. Trans. Missouri Acad. Sci. 5: 26-38.
EDIGER, R. and N. SANTAMARIA.
1971. *Ratibida columnifera* (Compositae) in California. Madroño 21: 12.
EPLING, C.
1925. Monograph of the genus *Monardella*. Ann. Missouri Bot. Gard. 12: 1-106.

1938.   The Californian salvias. A review of *Salvia*, section *Audibertia*. Ann. Missouri Bot. Gard. 25: 95-188.

1947a. Natural hybridization of *Salvia apiana* and *S. mellifera*. Evolution 1: 69-78.

1947b. Actual and potential gene flow in natural populations. Amer. Nat. 81: 104-113.

1948.   A synopsis of the tribe Lepechineae. Brittonia 6: 352-364.

EPLING, C. and T. DOBZHANSKY.

1942.   Genetics of natural populations. VI. Microgeographical races in *Linanthus parryae*. Genetics 27: 317-332.

EPLING, C., H. LEWIS, and F. M. BALL.

1960.   The breeding group and seed storage: a study in population dynamics. Evolution 14: 238-255.

EPLING, C., H. LEWIS, and P. H. RAVEN.

1962.   Chromosomes of *Salvia*: section *Audibertia*. Aliso 5: 217-221.

ERNST, W. R.

1962.   The genera of Papaveraceae and Fumariaceae in the southeastern United States. Jour. Arnold Arb. 43: 315-343.

EVITT, W. R. and S. T. PIERCE.

1975.   Early Tertiary ages from the coastal belt of the Franciscan complex, northern California. Geology 3: 433-436.

FAVARGER, C. and J. CONTANDRIOPOULOS.

1961.   Essai sur l'endémisme. Bul. Soc. Bot. Suisse 71: 384-408.

FELLOWS, C. E.

1976.   Chromosome counts and a new combination in *Claytonia* sect. *Limnia* (Portulacaceae). Madroño 23: 296-297.

FERNALD, M. L.

1950.   Gray's Manual of Botany. Ed. 8. American Book, New York. lxiv + 1632 pp.

FEUER, S. and A. S. TOMB.

1977.   Pollen morphology and detailed structure of family Compositae, tribe Cichorieae. II. Subtribe Microseridinae. Amer. Jour. Bot. 64: 230-245.

FOSBERG, F. R.

1948.   Derivation of the flora of the Hawaiian Islands. *In* E. C. Zimmermann, Insects of Hawaii, Vol. 1, p. 107-119. University of Hawaii Press, Honolulu.

FOWELLS, H. A.

1965.   Silvics of forest trees of the United States. U.S. Dept. Agric. Forest Service, Agriculture Handbook 271. 762 pp.

FRENKEL, R. E.

1970.   Ruderal vegetation along some California roadsides. Univ. Calif. Publ. Geogr. 20: i-vii, 1-163.

FRYXELL, P. A.

1971.   A revision of *Phrymosia*. Madroño 21: 153-174.

1974.   The North American malvellas. Southw. Nat. 19: 97-103.

GALLOWAY, L. A.

1976.   Systematics of the North American desert species of *Abronia* and *Tripterocalyx* (Nyctaginaceae). Brittonia 27: 328-347.

GANKIN, R. and J. MAJOR.

1964.   *Arctostaphylos myrtifolia*, its biology and relationship to the problem of endemism. Ecology 45: 792-808.

GARBER, E. D.

1960.   The genus *Collinsia*. IX. Speciation and chromosome repatterning. Cytologia 25: 233-243.

GARBER, E. D. and S. L. BELL.

1962.   The genus *Collinsia*. XV. A cytogenetic study of three fertile interspecific hybrids with an interchange complex of six chromosomes. Bot. Gaz. 123: 190-197.

GARBER, E. D. and T. S. DHILLON.

1962.   The genus *Collinsia*. XX. Cytogenetic studies of interspecific hybrids. Bot. Gaz. 123: 292-298.

GARBER, E. D. and J. GORSIC.

1956.   The genus *Collinsia*. II. Interspecific hybrids involving *C. heterophylla*, *C. concolor*, and *C. sparsifolia*. Bot. Gaz. 118: 73-77.

GASTIL, R. G. and W. JENSKY.
   1973.   Evidence for strike-slip displacement beneath the Trans-Mexican volcanic belt. Stanford Univ. Publ. Geol. Sci. 13: 181-190.
GIANNASI, D. E. and T.-I. CHUANG.
   1976.   Flavonoid systematics of the genus *Perideridia* (Umbelliferae). Brittonia 28: 177-194.
GIBSON, D. N.
   1972.   Studies in American plants, III. Fieldiana Bot. 34: 57-88.
GILLETT, G. B.
   1955.   Variation and genetic relationships in the *Whitlavia* and *Gymnobythus* phacelias. Univ. Calif. Publ. Bot. 28: 19-78, pl. 3-5.
GILLETT, J. J.
   1976.   A new species of *Trifolium* (Leguminosae) from Baja California, Mexico. Madroño 23: 334-337.
GILLIS, W. T.
   1971.   The systematics and ecology of poison-ivy and the poison-oak (*Toxicodendron*, Anacardiaceae). Rhodora 73: 72-159, 161-237, 370-443, 465-540.
GODLEY, E. J.
   1975.   IV. Flora and vegetation. *In* Kuschel, G. (ed.), Biogeography and Ecology in New Zealand, p. 177-229. Dr. W. Junk b. v. Publ., The Hague.
GOODSPEED, T. H.
   1954.   The genus *Nicotiana*. Chron. Bot. 16: 1-536.
GORDON, T. R. and M. H. GRAYUM.
   1976.   *Sedum spathulifolium* (Crassulaceae), new to the Santa Monica Mountains, California. Madroño 23: 454.
GOTTLIEB, L. D.
   1968.   Hybridization between *Arctostaphylos viscida* and *A. canescens* in Oregon. Brittonia 20: 83-93.
   1971.   Evolutionary relationships in the outcrossing diploid species of *Stephanomeria* (Compositae). Evolution 25: 312-329.
   1972.   A proposal for classification of the annual species of *Stephanomeria* (Compositae). Madroño 21: 463-481.
   1973a.   Genetic differentiation, sympatric speciation and the origin of a diploid species of *Stephanomeria*. Amer. Jour. Bot. 60: 545-554.
   1973b.   Enzyme differentiation and phylogeny in *Clarkia franciscana, C. rubicunda* and *C. amoena*. Evolution 27: 204-214.
   1974a.   Genetic confirmation of the origin of *Clarkia lingulata*. Evolution 28: 244-250.
   1974b.   Genetic stability in a peripheral isolate of *Stephanomeria exigua* ssp. *coronaria* that fluctuates in population size. Genetics 76: 551-556.
   1975.   Allelic diversity in the outcrossing annual plant *Stephanomeria exigua* ssp. *carotifera* (Compositae). Evolution 29: 213-225.
GOULD, F. W.
   1974.   Nomenclatural changes in the Poaceae. Brittonia 26: 59-60.
GRANT, K. A. and V. GRANT.
   1964.   Mechanical isolation of *Salvia apiana* and *Salvia mellifera* (Labiatae). Evolution 18: 196-212.
GRANT, V.
   1953.   The role of hybridization in the evolution of the leafy-stemmed gilias. Evolution 7: 51-64.
   1959.   Natural History of the Phlox Family. Martinus Nijhoff, The Hague. xv + 280 pp.
   1971.   Plant Speciation. Columbia Univ. Press, New York and London. x + 435 pp.
GRANT, W. F. and B. S. SIDHU.
   1967.   Basic chromosome number, cyanogenetic glucoside variation, and geographic distribution of *Lotus* species. Canad. Jour. Bot. 45: 639-647.
GRAY, A.
   1859.   Memoir on the botany of Japan, in its relation to that of North America, and of other parts of the northern temperate zone. Mem. Amer. Acad. II. 6: 1-66.
   1878.   Forest geography and archaeology. Amer. Jour. Sci. & Arts III. 16: 85-95.

1884.   Characteristics of the North American flora. Amer. Jour. Sci. & Arts III. 28: 323-340.

GRIFFIN, J. R.
1966.   Notes on disjunct foothill species near Burney, California. Leafl. West. Bot. 10: 296-298.
1975.   Plants of the highest Santa Lucia and Diablo Range peaks, California. U.S. Dept. Agric. Serv. Res. Pap. PSW-110: 1-50.

GRIFFIN, J. R. and W. B. CRITCHFIELD.
1972.   The distribution of forest trees in California. U.S. Dept. Agric. Forest Service Research Paper PSW-82: 1-114.

GRIFFIN, J. R. and C. O. STONE.
1967.   MacNab cypress in northern California: a geographic review. Madroño 19: 19-27.

HACKEL, O.
1966.   Summary of the geology of the Great Valley. Calif. Div. Mines and Geology Bull. 190: 217-238.

HALLER, J. R.
1961.   Some recent observations on ponderosa, Jeffrey and Washoe pines in northeastern California. Madroño 16: 126-132.
1962.   Variation and hybridization in ponderosa and Jeffrey pines. Univ. Calif. Publ. Bot. 34: 123-166.

HAMRICK, J. L.
1976.   Variation and selection in western montane species II. Variation within and between populations of white fir on an elevational transect. Theoret. Appl. Genet. 47: 27-34.

HAMRICK, J. L. and R. W. ALLARD.
1972.   Microgeographical variation in allozyme frequencies in *Avena barbata*. Proc. Natl. Acad. Sci. U.S. 69: 2100-2104.
1975.   Correlations between quantitative characters and enzyme genotypes in *Avena barbata*. Evolution 29: 438-442.

HAMRICK, J. L. and W. J. LIBBY.
1972.   Variation and selection in western U.S. montane species. Silvae Genet. 21: 29-35.

HARDHAM, C. G.
1962.   The Santa Lucia *Cupressus sargentii* groves and their associated northern hydrophilous and endemic species. Madroño 16: 173-204.
1966a. Three diploid species of the *Monardella villosa* complex. Leafl. West. Bot. 10: 237-242.
1966b. Two more diploid segregates of the *Monardella villosa* complex. Leafl. West. Bot. 10: 320-326.

HARDIN, J. W.
1957.   A revision of the American Hippocastanaceae. I. Brittonia 9: 145-171. II. Brittonia 9: 173-195.

HARDING, J. and C. B. MANKINEN.
1967.   Genetics of *Lupinus*. I. Variations in flower color from natural populations of *Lupinus nanus*. Canad. Jour. Bot. 45: 1831-1836.
1971.   Genetics of *Lupinus*. III. Evidence for genetic differentiation and colonization in *Lupinus succulentus*. Madroño 21: 222-235.

HARMON, W. E. and D. B. DUNN.
1968.   Experimental studies of compatibility of the isolated endemic, *Lupinus niveus*. Trans. Missouri Acad. Sci. 2: 84-90.

HASTINGS, J. R., R. M. TURNER, and D. K. WARREN.
1972.   An atlas on some plant distributions in the Sonoran Desert. Univ. Arizona Inst. Atmos. Phys. Tech. Rept. Meteorol. Climatol. Arid Reg. 21: i-xiii, 1-255.

HAWKSWORTH, F. G. and D. WIENS.
1972.   Biology and classification of dwarf mistletoes (*Arceuthobium*). U.S. Dept. Agric. Handbook 401: i-viii, 1-234.

HECKARD, L. R.
1968.   Chromosome numbers and polyploidy in *Castilleja* (Scrophulariaceae). Brittonia 20: 212-226.
1969.   A new *Campanula* from northern California. Madroño 20: 231-235.
1973.   A taxonomic re-interpretation of the *Orobanche californica* complex. Madroño 22: 41-70.

HECKARD, L. R. and R. BACIGALUPI.
  1970.   A new species of *Castilleja* from the southern Sierra Nevada. Madroño 20: 209-213.
HECKARD, L. R. and G. L. STEBBINS.
  1974.   A new *Lewisia* (Portulacaceae) from the Sierra Nevada of California. Brittonia 26: 305-308.
HEDBERG, O.
  1946.   Pollen morphology in the genus *Polygonum* L. s. lat. and its taxonomical significance. Svensk. Bot. Tidskr. 40: 371-404.
HEISER, C. B., JR.
  1949.   Study in the evolution of the sunflower species *Helianthus annuus* and *H. bolanderi*. Univ. Calif. Publ. Bot. 23: 157-208.
HENDERSON, D. M.
  1976.   A biosystematic study of Pacific Northwestern blue-eyed grasses (*Sisyrinchium*, Iridaceae). Brittonia 28: 149-176.
HENDRY, G. W.
  1931.   The adobe brick as an historical source. Agric. Hist. 5: 110-127.
HENRICKSON, J.
  1972.   A taxonomic revision of the Fouquieriaceae. Aliso 7: 439-537.
HENRICKSON, J. and B. PRIGGE.
  1975.   White fir in the mountains of eastern Mohave Desert of California. Madroño 23: 164-168.
HERMANN, F. J.
  1948.   The *Juncus triformis* group in North America. Leafl. West. Bot. 5: 109-124.
HIESEY, W. M., M. A. NOBS, and O. BJÖRKMAN.
  1971.   Experimental studies on the nature of species. V. Biosystematic genetics, and physiological ecology of the *Erythranthe* section of *Mimulus*. Carnegie Inst. Wash. Pub. 628: i-vi, 1-213.
HILL, A. J.
  1973.   A distinctive new *Calochortus* (Liliaceae) from Marin County, California. Madroño 22: 100-104.
HILLEBRAND, W.
  1888.   Flora of the Hawaiian Islands. Privately published, Heidelberg. Reprinted 1968 by Hafner, New York, xcvi + 673 pp.
HINTON, W. F.
  1975.   Systematics of the *Calyptridium umbellatum* complex (Portulacaceae). Brittonia 27: 197-208.
  1976a.   Introgression and the evolution of selfing in *Calyptridium monospermum* (Portulacaceae). Syst. Bot. 1: 85-90.
  1976b.   The evolution of insect-mediated self-pollination from an outcrossing system in *Calyptridium* (Portulacaceae). Amer. Jour. Bot. 63: 979-986.
HITCHCOCK, C. L.
  1952.   A revision of the North American species of *Lathyrus*. Univ. Wash. Publ. Biol. 15: 1-104.
  1957.   A study of the perennial species of *Sidalcea*. Part I. Taxonomy. Univ. Wash. Publ. Biol. 18: 1-82.
HITCHCOCK, C. L. and B. MAGUIRE.
  1947.   A revision of the North American species of *Silene*. Univ. Wash. Publ. Biol. 13: 1-73.
HOFFMAN, F. W.
  1952.   Studies in *Streptanthus*. A new *Streptanthus* complex in California. Madroño 11: 221-233.
HOLM, R. W.
  1952.   The American species of *Sarcostemma*. Ann. Missouri Bot. Gard. 37: 477-560.
HOLMGREN, A. H., L. M. SHULTZ, and T. K. LOWREY.
  1976.   *Sphaeromeria*, a genus closer to *Artemisia* than to *Tanacetum* (Asteraceae: Anthemideae). Brittonia 28: 252-262.
HOLUB, J.
  1976.   Portion of Asteraceae-Inuleae. *In* T. G. Tutin et al. (eds.), Flora Europaea 4: 121-128. Cambridge Univ. Press, Cambridge, England.
HOOVER, R. F.
  1940.   The genus *Dichelostemma*. Amer. Midl. Nat. 24: 463-476.
  1941.   A systematic study of *Triteleia*. Amer. Midl. Nat. 25: 73-100.

1955.   Further observations on *Brodiaea* and some related genera. Pl. Life (Herbertia ed.) 11: 13-23.

1970.   The Vascular Plants of San Luis Obispo County, California. Univ. California Press, Berkeley, Los Angeles and London. 350 pp.

n.d.   Endemism in the Flora of the Great Valley of California. Ph.D. Dissertation, Univ. California, Berkeley. 76 pp.

HOWELL, J. T.

1957.   The California flora and its province. Leafl. West. Bot. 8: 133-138.

1960.   The endemic pteridophytes of the California Floral Province. Amer. Fern Jour. 50: 15-25.

1970.   Marin Flora. Ed. 2. Univ. California Press, Berkeley, Los Angeles and London. ix + 366 pp.

1971.   A new name for "winter fat." Wasmann Jour. Biol. 29: 105.

1972.   A statistical estimate of Munz' Supplement to *A California Flora*. Wasmann Jour. Biol. 30: 93-96.

HOWELL, J. T., P. H. RAVEN, and P. RUBTZOFF.

1958.   A flora of San Francisco, California. Wasmann Jour. Biol. 16: 1-157.

HOWITT, B. F. and J. T. HOWELL.

1964.   The vascular plants of Monterey County, California. Wasmann Jour. Biol. 22 (1): 1-184.

HSIAO, J.-Y.

1973.   A numerical taxonomic study of the genus *Platanus* based on morphological and phenolic characters. Amer. Jour. Bot. 60: 678-684.

HUBER, H.

1969.   Die Samenmerkmale und Verwandtschaftsverhältnisse der Liliifloren. Mitt. Bot. Staatsamml. München 8: 219-538.

HUETHER, C. A., Jr.

1966.   The extent of variability for a canalized character (corolla lobe number) in natural populations of *Linanthus* (Benth.). Ph.D. Dissertation, University of California, Davis.

1968.   Exposure of natural genetic variability underlying the pentamerous corolla constance in *Linanthus androsaceus* ssp. *androsaceus*. Genetics 60: 123-146.

1969.   Constancy of the pentamerous corolla phenotype in natural populations of *Linanthus*. Evolution 23: 572-588.

HULTÉN, E.

1968.   Flora of Alaska and Neighboring Territories. Stanford Univ. Press, Stanford, Calif. xxii + 1008 pp.

HUTCHINSON, J.

1973.   The Families of Flowering Plants. Ed. 3. Clarendon Press, Oxford. xix + 968 pp.

ILTIS, H. H.

1957.   Studies in the Capparidaceae. III. Evolution and phylogeny of the western North American Cleomoideae. Ann. Missouri Bot. Gard. 44: 77-119.

IMAM, A. G. and R. W. ALLARD.

1965.   Population studies in predominantly self-pollinated species. VI. Genetic variability between and within natural populations of wild oats, *Avena fatua* L., from differing habitats in California. Genetics 51: 49-62.

JAIN, S. K.

1969.   Comparative ecogenetics of two *Avena* species occurring in central California. Evol. Biol. 3: 73-118.

JAIN, S. K. (ed.).

1976.   Vernal pools. Their ecology and conservation. Univ. Calif. Davis Inst. Ecol. Publ. 9: i-vi, 1-93.

JAIN, S. K. and D. R. MARSHALL.

1967.   Population studies in predominantly self-pollinating species. X. Variation in natural populations of *Avena fatua* and *A. barbata*. Amer. Nat. 101: 19-33.

JAIN, S. K., D. R. MARSHALL, and K. WU.

1970.   Genetic variability in natural populations of softchess (*Bromus mollis* L.). Evolution 24: 649-659.

JENNY, H., R. J. ARKLEY, and A. M. SCHULTZ.

1969.   The pygmy forest-podsol ecosystem and its dune associates of the Mendocino coast. Madroño 20: 60-74.

JEPSON, W. L.
  1925. A Manual of the Flowering Plants of California. Univ. California Press, Berkeley. 1238 pp.
JOHNSON, M. P. and P. H. RAVEN.
  1973. Species number and endemism: The Galápagos Archipelago revisited. Science 179: 893-895.
JOHNSON, N. K.
  1972. Origin and differentiation of the airfauna of the Channel Islands, California. Condor 74: 295-315.
  1975. Controls of number of bird species on montane islands in the Great Basin. Evolution 29: 545-567.
JOHNSTON, M. C.
  1971. Revision of *Colubrina* (Rhamnaceae). Brittonia 23: 2-53.
KAM, Y. K. and J. MAZE.
  1974. Studies on the relationships and evolution of supraspecific taxa utilizing developmental data. II. Relationships and evolution of *Oryzopsis hymenoides, O. virescens, O. kingii, O. micrantha,* and *O. asperifolia.* Bot. Gaz. 135: 227-247.
KANNENBERG, L. W. and R. W. ALLARD.
  1967. Population studies in predominantly self-pollinating species. VIII. Genetic variability in the *Festuca microstachys* complex. Evolution 21: 227-240.
KARIG, D. E. and JENSKY, W.
  1972. The proto-Gulf of California. Earth and Planetary Sci. Letters 17: 169-174.
KEATOR, R.
  1968. A taxonomic and ecological study of the genus *Dichelostemma* (Amaryllidaceae). Ph.D. Dissertation, Univ. California, Berkeley. xiii + 467 pp.
KEELEY, J. E.
  1975. Longevity of nonsprouting *Ceanothus.* Amer. Midl. Nat. 93: 504-507.
  1976. Morphological evidence of hybridization between *Arctostaphylos glauca* and *A. pungens* (Ericaceae). Madroño 23: 427-434.
KLEIN, W. M.
  1970. The evolution of three diploid species of *Oenothera* subgenus *Anogra.* Evolution 24: 578-597.
KNOWLES, P. F.
  1943. Improving an annual brome grass, *Bromus mollis* L., for range purposes. Jour. Amer. Soc. Agron. 35: 584-594.
KRUCKEBERG, A. R.
  1951. Intraspecific variation in the response of certain native plant species to serpentine soil. Amer. Jour. Bot. 38: 408-419.
  1954. The plant species in relation to serpentine soils. Ecology 33: 267-274.
  1955. Interspecific hybridization of *Silene.* Amer. Jour. Bot. 42: 373-378.
  1957. Variation in fertility of hybrids between isolated populations of the serpentine species, *Streptanthus glandulosus* Hook. Evolution 11: 185-211.
  1958. The taxonomy of the species complex, *Streptanthus glandulosus* Hook. Madroño 14: 217-227.
  1961. Artificial crosses of western North American silenes. Brittonia 13: 305-333.
  1969. Soil diversity and the distribution of plants, with examples from western North America. Madroño 20: 129-154.
KYHOS, D. W.
  1965. The independent aneuploid origin of two species of *Chaenactis* (Compositae) from a common ancestor. Evolution 19: 26-43.
LANNER, R. M.
  1974. A new pine from Baja California and the hybrid origin of *Pinus quadrifolia.* Southw. Nat. 19: 75-95.
LAWRENCE, L., R. BARTSCHOT, E. ZAVARIN, and J. R. GRIFFIN.
  1975. Natural hybridization of *Cupressus sargentii* and *C. macnabiana* and the composition of the derived essential oils. Biochem. Syst. Ecol. 2: 113-119.

LENZ, L. W.

1958.   A revision of the Pacific Coast irises. Aliso 4: 1-72.

1971.   Two new species of *Dandya* (Liliaceae) from Mexico and a reexamination of *Bessera* and *Behria*. Aliso 7: 313-320.

1974.   A new species of *Dichelostemma* (Liliaceae) from California. Aliso 8: 129-131.

1975.   A biosystematic study of *Triteleia* (Liliaceae). I. Revision of the species of section *Calliprora*. Aliso 8: 221-258.

LESKINEN, P. H.

1975.   Occurrence of oaks in Late Pleistocene vegetation in the Mojave Desert of Nevada. Madroño 23: 234-235.

LEWIS, H.

1945.   A revision of the genus *Trichostema*. Brittonia 5: 276-303.

1953a. The mechanism of evolution in the genus *Clarkia*. Evolution 7: 1-20.

1953b. Chromosome phylogeny and habitat preference in *Clarkia*. Evolution 7: 102-109.

1962.   Catastrophic selection as a factor in speciation. Evolution 16: 257-271.

1966.   Speciation in flowering plants. Science 152: 167-172.

1969a. Speciation. Taxon 18: 21-25.

1969b. Evolutionary processes in the ecosystem. BioScience 19: 223-227.

1972.   The origin of endemics in the California flora. *In* Valentine, D. H. (ed.), Taxonomy, Phytogeography and Evolution, pp. 179-189. Academic Press, London and New York.

1973.   The origin of diploid neospecies in *Clarkia*. Amer. Nat. 107: 161-170.

LEWIS, H. and C. EPLING.

1940.   Three species pairs from southern and Lower California. Amer. Midl. Nat. 24: 743-749.

1954.   A taxonomic study of California delphiniums. Brittonia 8: 1-22.

1959.   *Delphinium gypsophilum*, a diploid species of hybrid origin. Evolution 13: 511-525.

LEWIS, H. and M. E. LEWIS.

1955.   The genus *Clarkia*. Univ. Calif. Publ. Bot. 20: 241-392.

LEWIS, H. and P. H. RAVEN.

1958.   Rapid evolution in *Clarkia*. Evolution 12: 319-336.

LEWIS, H. and M. R. ROBERTS.

1956.   The origin of *Clarkia lingulata*. Evolution 10: 126-138.

LIBBY, W. J.

1958.   The backcross hybrid Jeffrey X (Jeffrey X Coulter) pine. Jour. Forest. 56: 840-842.

LINDSAY, G.

1963.   The genus *Lophocereus*. Cact. Succ. Jour. 35: 176-192.

LINHART, Y. B.

1975.   Evolutionary studies of plant populations in vernal pools. Univ. Calif. Davis Inst. Ecol. Publ. 9: 40-46.

LITTLE, E. L., Jr.

1970.   Names of New World cypresses (*Cupressus*). Phytologia 20: 429-445.

LITTLE, E. L., JR. and W. B. CRITCHFIELD.

1969.   Subdivisions of the genus *Pinus* (pines). U.S. Dept. Agr. For. Serv. Misc. Publ. 1144: i-iv, 1-51.

LLOYD, R. M. and R. S. MITCHELL.

1973.   A Flora of the White Mountains of California and Nevada. Univ. California Press, Berkeley. 202 pp.

LONARD, R. I. and F. W. GOULD.

1974.   The North American species of *Vulpia* (Gramineae). Madroño 22: 217-230.

MAJOR, J.

1975.   Laymen's botany in California. Ecology 56: 1020-1021.

MAJOR, J. and S. A. BAMBERG.

1963.   Some Cordilleran plant species new for the Sierra Nevada of California. Madroño 17: 93-109.

1967.   Some Cordilleran plants disjunct in the Sierra Nevada of California, and their bearing on Pleistocene ecological conditions. *In* H. E. Wright and W. H. Osborn (ed.), Arctic and Alpine Environments, pp. 171-189. Indiana Univ. Press, Bloomington, Ind.

MASON, C. T., JR.

1975.   *Apacheria chiricahuensis*: A new genus and species from Arizona. Madroño 23: 105-108.

MASON, H. L.
  1934.   Pleistocene flora of the Tomales Formation. Carnegie Inst. Wash. Pub. 415: 81-179.
  1946a. The edaphic factor in narrow endemism. I. The nature of environmental influences. Madroño 8: 209-226.
  1946b. The edaphic factor in narrow endemism. II. The geographic occurrence of plants of highly restricted patterns of distribution. Madroño 8: 241-257.
MATHIAS, M. E. and L. CONSTANCE.
  1965.   A revision of the genus *Bowlesia* Ruiz & Pav. (Umbelliferae-Hydrocotyloideae) and its relatives. Univ. Calif. Publ. Bot. 38: 1-73.
  1977.   Two new local Umbelliferae (Apiaceae) from California. Madroño 24: 78-83.
McMILLAN, C.
  1956.   The edaphic restriction of *Cupressus* and *Pinus* in the Coast Ranges of central California. Ecol. Monogr. 26: 177-212.
  1959.   Survival of transplanted *Cupressus* in the pygmy forests of Mendocino County, California. Madroño 15: 1-4.
  1964.   Survival of transplanted *Cupressus* and *Pinus* after thirteen years in Mendocino County, California. Madroño 17: 250-253.
McMINN, H. E.
  1951.   An Illustrated Manual of California Shrubs. Univ. California Press, Berkeley and Los Angeles. xi + 663 pp.
McNEILL, J.
  1975.   A generic revision of Portulacaceae tribe Montieae using techniques of numerical taxonomy. Canad. Jour. Bot. 53: 789-809.
MERTENS, T. R. and P. H. RAVEN.
  1965.   Taxonomy of *Polygonum*, section *Polygonum* (*Avicularia*) in North America. Madroño 18: 85-92.
MEUSEL, H.
  1969.   Beziehungen in der Florendifferenzierung von Eurasien und Nordamerika. Flora, Abt. B. 158: 537-564.
MILLER, G. N.
  1955.   *Fraxinus* in North America. Cornell Univ. Agric. Expt. Sta. Mem. 335.
MILLER, J. H.
  1976.   Variation in populations of *Claytonia perfoliata* (Portulacaceae). Syst. Bot. 1: 20-34.
MOORE, D. G.
  1973.   Plate-edge deformation and crustal growth, Gulf of California structural province. Geol. Soc. Amer. Bull. 84: 1883-1906.
MOORE, D. M. and H. LEWIS.
  1965.   The evolution of self-pollination in *Clarkia xantiana*. Evolution 19: 104-114.
MOORE, D. M. and P. H. RAVEN.
  1970.   Cytogenetics, distribution, and amphitropical affinities of South American *Camissonia* (Onagraceae). Evolution 24: 816-823.
MOORE, H. E., JR.
  1953.   The genus *Milla* (Amaryllidaceae-Allieae) and its allies. Gentes Herb. 8: 263-294.
  1973.   The major groups of palms and their distribution. Gentes Herb. 11: 27-140.
  1975.   Nomenclatural notes for *Hortus Third*. Baileya 19: 163-171.
MOORING, J. S.
  1965.   Chromosome studies in *Chaenactis* and *Chamaechaenactis* (Compositae, Helenieae). Brittonia 17: 17-25.
  1973.   Chromosome counts in *Eriophyllum* and other Helenieae (Compositae). Madroño 22: 95-97.
  1975.   A cytogeographic study of *Eriophyllum lanatum* (Compositae, Helenieae). Amer. Jour. Bot. 62: 1027-1037.
MORAN, R.
  1950a. *Mesembryanthemum* in California. Madroño 10: 161-163.
  1950b. Whence *Sedum pinetorum* Brandegee? Leafl. West. Bot. 6: 62-64.
  1972.   Las plantas vasculares de la Isla de Cedros. Calafia 2(1): 34-35.
  1973.   *Ornithostaphylos* (Ericaceae) in California. Madroño 22: 214.

MOREY, D. H.
1959. Changes in nomenclature in the genus *Plectritis*. Contr. Dudley Herb. 5: 119-121.
MOSQUIN, T.
1962. *Clarkia stellata*, a new species from California. Leafl. West. Bot. 9: 215-216.
1964. Chromosomal repatterning in *Clarkia rhomboidea* as evidence for post-Pleistocene changes in distribution. Evolution 18: 12-15.
1970. Chromosome numbers and a proposal for classification in *Sisyrinchium* (Iridaceae). Madroño 20: 269-275.
MUIR, J.
1912. The Yosemite. Appleton-Century, New York.
MUÑOZ, P. C.
1966. Sinopsis de la flora chilena. Ed. 1. Univ. de Chile. 500 pp.
MUNZ, P. A.
1935. A Manual of Southern California Botany. Claremont College, Claremont, Calif. xxxix + 642 pp.
1968. Supplement to *A California Flora*. Univ. California Press, Berkeley and Los Angeles. 224 pp.
1969. California miscellany—VII. Aliso 7: 65-71.
1974. A Flora of Southern California. Univ. California Press, Berkeley, Los Angeles, London. 1086 pp.
MUNZ, P. A. and D. D. KECK.
1959. A California Flora. Univ. California Press, Berkeley and Los Angeles. 1681 pp.
NAVEH, Z.
1967. Mediterranean ecosystems and vegetation types in California and Israel. Ecology 48: 445-459.
NELSON, A. P.
1965. Taxonomic and evolutionary implications of lawn races in *Prunella vulgaris* (Labiatae). Brittonia 17: 160-174.
NELSON, C. H. and G. S. VAN HORN.
1976. A new simplified method for constructing Wagner networks and the cladistics of *Pentachaeta* (Compositae, Astereae). Brittonia 27: 362-372.
NIEHAUS, T. F.
1971. A biosystematic study of the genus *Brodiaea*. Univ. Calif. Publ. Bot. 60: 1-66, 1 pl.
NOBS, M. A.
1963. Experimental studies on species relationships in *Ceanothus*. Carnegie Inst. Wash. Publ. 623: 1-94.
NOLDECKE, A. M. and J. T. HOWELL.
1960. Endemism and *A California Flora*. Leafl. West. Bot. 9: 124-127.
NOWACKI, E. and D. B. DUNN.
1964. Shrubby California lupines and relationships suggested by alkaloid content. Genet. Polon. 5: 47-56.
OHWI, J.
1965. Flora of Japan. Smithsonian Institution, Washington, D.C. 1067 pp.
ORNDUFF, R.
1963. Experimental studies in two genera of Helenieae (Compositae): *Blennosperma* and *Lasthenia*. Quart. Rev. Biol. 38: 141-150.
1964. Biosystematics of *Blennosperma*. Brittonia 16: 289-295.
1965. Ornithocoprophilous endemism in Pacific Basin angiosperms. Ecology 46: 864-867.
1966. A biosystematic survey of the goldfield genus *Lasthenia* (Compositae: Helenieae). Univ. Calif. Publ. Bot. 40: 1-92.
1969a. *Limnanthes vinculans*, a new California endemic. Brittonia 21: 11-14.
1969b. The origin and relationships of *Lasthenia burkei* (Compositae). Amer. Jour. Bot. 56: 1042-1047.
1969c. Ecology, morphology, and systematics of *Jepsonia* (Saxifragaceae). Brittonia 21: 286-298.
1971. Systematic studies of Limnanthaceae. Madroño 21: 103-111.
1976. Speciation and oligogenic differentiation in *Lasthenia* (Compositae). Syst. Bot. 1: 91-96.

ORNDUFF, R. and B. A. BOHM.
1975. Relationships of *Tracyina* and *Rigiopappus* (Compositae). Madroño 23: 53-55.
ORNDUFF, R. and T. CROVELLO.
1968. Numerical taxonomy of Limnanthaceae. Amer. Jour. Bot. 55: 173-182.
ORNDUFF, R., N. A. M. SALEH, and B. A. BOHM.
1973. The flavonoids and affinities of *Blennosperma* and *Crocidium*. Taxon 22: 407-412.
ORSHAN, G.
1953. Note on the application of Raunkiaer's system of life forms in arid regions. Palest. Jour. Bot. 6: 120-122.
OWNBEY, G. B.
1947. Monograph of the North American species of *Corydalis*. Ann. Missouri Bot. Gard. 34: 187-259.
OWNBEY, G. B., P. H. RAVEN, and D. W. KYHOS.
1976. Chromosome numbers in some North American species of the genus *Cirsium*. III. Western United States, Mexico, and Guatemala. Brittonia 27: 297-304.
PANETSOS, C. and H. G. BAKER.
1968. The origin of variation in "wild" *Raphanus sativus* (Cruciferae) in California. Genetics 38: 243-274.
PARISH, S. B.
1920. The immigrant plants of Southern California. So. Calif. Acad. Sci. Bull. 14 (4): 3-30.
PARKER, W. H.
1977. Comparison of numerical taxonomic methods used to estimate flavonoid similarities in the Limnanthaceae. Brittonia 28: 390-399.
PAYNE, W. W.
1964. A re-evaluation of the genus *Ambrosia* (Compositae). Jour. Arnold Arb. 45: 401-438.
PAYNE, W. W., T. A. GEISSMAN, A. J. LUCAS, and T. SAITOH.
1973. Chemosystematics and taxonomy of *Ambrosia chamissonis*. Biochem. Syst. 1: 21-33.
PECK, M. E.
1961. A Manual of the Higher Plants of Oregon. Ed. 2. Bindfords & Mort, Publishers, with Oregon State Univ. Press, Corvallis. 936 pp.
PERRY, L. M.
1931. Concerning a Californian *Convolvulus*. Rhodora 33: 76-77.
PICHI-SERMOLLI, R.
1950. Sulla sistematica e nomenclatura di alcune piante dell'Abissinia. Webbia 7: 325-351.
PIEHL, M. A.
1965. The natural history and taxonomy of *Comandra* (Santalaceae). Mem. Torrey Bot. Club 22: 1-97.
PILZ, G. E.
1974. Systematics of *Mirabilis* subgenus *Quamoclidion* (Nyctaginaceae). Ph.D. Dissertation, Univ. California, Berkeley.
PLATT, J. P.
1976. The petrology, structure, and geologic history of the Catalina schist terrain, southern California. Univ. Calif. Publ. Geol. Sci. 112: 1-111.
PORTER, D. M.
1976. Geography and dispersal of Galápagos Islands vascular plants. Nature 264: 745-746.
POWELL, A. M., D. W. KYHOS, and P. H. RAVEN.
1974. Chromosome numbers in Compositae. X. Amer. Jour. Bot. 61: 909-913.
POWELL, W. R.
1974. Inventory of rare and endangered vascular plants of California. Calif. Native Plant Soc. Spec. Publ. 1: i-iii, 1-56.
PROCTOR, J. and S. R. J. WOODELL.
1975. The ecology of serpentine soils. Adv. in Ecol. Research 9: 255-366.
RADFORD, A. E., H. E. AHLES, and C. R. BELL.
1968. Manual of the Vascular Flora of the Carolinas. Univ. North Carolina Press, Chapel Hill. lxi + 1183 pp.

RAUNKIAER, C.
1934.   The Life Forms of Plants and Statistical Plant Geography. Clarendon Press, Oxford. 632 pp.

RAVEN, P. H.
1962.   The systematics of *Oenothera* subgenus *Chylismia*. Univ. Calif. Publ. Bot. 34: 1-122.
1963.   A flora of San Clemente Island, California. Aliso 5: 289-347.
1964.   Catastrophic selection and edaphic endemism. Evolution 18: 336-338.
1967.   The floristics of the California Islands. *In* Philbrick, R. (ed.), Proceedings of the Symposium on the Biology of the California Islands, p. 57-67. Santa Barbara Botanic Garden.
1969.   A revision of the genus *Camissonia* (Onagraceae). Contr. U.S. Natl. Herb. 37: 161-396.
1971.   The relationship between "mediterranean" floras. *In* Davis, P. H., P. C. Harper, and I. C. Hedge (eds.), Plant Life of South-West Asia, p. 119-134. Univ. Press, Aberdeen.
1972.   Plant species disjunctions: A summary. Ann. Missouri Bot. Gard. 59: 234-246.
1973.   The evolution of "mediterranean" floras. *In* H. Mooney and F. di Castri (eds.), Evolution of Mediterranean Ecosystems, p. 213-224. Springer-Verlag, New York, Heidelberg, Berlin.
1976a. Generic and sectional delimitation in Onagraceae, tribe Epilobieae. Ann. Missouri Bot. Gard. 63: 326-340.
1976b. Systematics and plant population biology. Syst. Bot. 1: 284-316.
1977.   The California flora. *In* M. Barbour and J. Major (eds.), Terrestrial Vegetation of California, p. 109-137. John Wiley Interscience, New York.

RAVEN, P. H. and D. I. AXELROD.
1974.   Angiosperm biogeography and past continental movements. Ann. Missouri Bot. Gard. 61: 539-673.

RAVEN, P. H. and M. E. MATHIAS.
1960.   *Sanicula deserticola*, an endemic of Baja California. Madroño 15: 193-197.

RAVEN, P. H. and J. H. THOMAS.
1970.   *Iris pseudacorus* in western North America. Madroño 20: 390-391.

REEDER, J. R.
1965.   The tribe Orcuttieae and the subtribes of the Pappophoreae (Gramineae). Madroño 18: 18-28.

REVEAL, J. L.
1969.   The subgeneric concept in *Eriogonum* (Polygonaceae). *In* Gunckel, J. E. (ed.), Current Topics in Plant Science, p. 229-249. Academic Press, New York and London.
1970.   Additional notes on the California buckwheats (*Eriogonum*, Polygonaceae). Aliso 7: 217-230.
1971.   A new annual *Eriogonum* (Polygonaceae) from southern Nevada and adjacent California. Aliso 7: 357-360.
1972.   Two new species of *Eriogonum* (Polygonaceae) from California and adjacent states. Aliso 7: 415-419.
1977.   Distribution and phylogeny of Eriogonoideae (Polygonaceae). *In* Harper, K. T. and J. T. Reveal (eds.), Intermountain Biogeography. Memoirs of the Great Basin Naturalist. Brigham Young Univ., Provo, Utah (in press).

REVEAL, J. L. and B. J. ERTTER.
1977.   *Goodmania* (Polygonaceae), a new genus from California. Brittonia 28: 427-429.

REVEAL, J. L. and J. T. HOWELL.
1976.   *Dedeckera* (Polygonaceae), a new genus from California. Brittonia 28: 245-251.

REVEAL, J. L. and R. M. KING.
1973.   Re-establishment of *Acourtia* D. Don (Asteraceae). Phytologia 27: 228-232.

REVEAL, J. L. and P. A. MUNZ.
1968.   *Eriogonum. In* P. A. Munz, Supplement to *A California Flora*, p. 33-72. Univ. California Press, Berkeley.

RIPLEY, S. D.
1975.   Report on endangered and threatened plant species of the United States. Serial No. 94-A: i-iv, 1-200. U.S. Govt. Printing Office, Washington, D.C.

ROBBINS, W. W., M. K. BELLUE, and W. S. BALL.
1951.   Weeds of California. State of California, Department of Agriculture. 547 pp.

ROBERTSON, K. R.
1972.   The genera of Geraniaceae in the southeastern United States. Jour. Arnold Arb. 53: 182-201.

ROBERTY, G. and S. VAUTIER.
  1964.   Les genres de Polygonacées. Boissiera 10: 7-128.
RODMAN, J. E.
  1974.   Systematics and evolution of the genus *Cakile* (Cruciferae). Contr. Gray Herb. 205: 3-146.
ROGERS, C. M.
  1975.   Relationships of *Hesperolinon* and *Linum* (Linaceae). Madroño 23: 153-159.
ROLLINS, R. C.
  1947.   Generic revisions in the Cruciferae: *Sibara*. Contr. Gray Herb. 165: 133-143.
ROOF, J.
  1976.   A fresh approach to the genus *Arctostaphylos* in California. Four Seasons 5(6): 1-24.
RUNE, O.
  1954.   Notes on the flora of the Gaspé Peninsula. Svensk. Bot. Tidskr. 48: 117-136.
SHAN, R. H. and L. CONSTANCE.
  1951.   The genus *Sanicula* (Umbelliferae) in the Old World and the New. Univ. Calif. Publ. Bot. 25: 1-78.
SHARSMITH, C. W.
  1940.   A contribution to the history of the alpine flora of the Sierra Nevada. Ph.D. Thesis. Univ. California, Berkeley, Calif. 274 pp.
SHARSMITH, H.
  1961.   The genus *Hesperolinon* (Linaceae). Univ. Calif. Publ. Bot. 32: 235-314.
SHAW, R. J.
  1962.   The biosystematics of *Scrophularia* in western North America. Aliso 5: 147-178.
SHETLER, S. G. and H. R. MEADOW.
  1972.   Provisional checklist of species for Flora North America. FNA Report 64: 1-648.
SHREVE, F.
  1936.   The transition from desert to chaparral in Baja California. Madroño 3: 257-264.
  1942.   The desert vegetation of North America. Bot. Rev. 8: 195-246.
SHREVE, F. and I. M. WIGGINS.
  1964.   Vegetation and Flora of the Sonoran Desert. 2 vols. Stanford Univ. Press, Stanford, Calif. x + 1740 pp.
SIMBERLOFF, D. S.
  1970.   Taxonomic diversity of island biotas. Evolution 24: 23-47.
SKVARLA, J. J. and B. L. TURNER.
  1966.   Pollen wall ultrastructure and its bearing on the systematic position of *Blennosperma* and *Crocidium* (Compositae). Amer. Jour. Bot. 53: 555-563.
SMALL, E.
  1971a. The evolution of reproductive isolation in *Clarkia*, section *Myxocarpa*. Evolution 25: 330-346.
  1971b. The systematics of *Clarkia*, section *Myxocarpa*. Canad. Jour. Bot. 49: 1211-1217
SMILEY, F. J.
  1921.   A report on the boreal flora of the Sierra Nevada of California. Univ. Calif. Publ. Bot. 9: 1-423.
SMITH, C. F.
  1976.   A Flora of the Santa Barbara Region, California. Santa Barbara Museum of Natural History, Santa Barbara. iv + 331 pp.
SMITH, D. M., S. P. CRAIG, and J. SANTAROSA.
  1971.   Cytological and chemical variation in *Pityrogramma triangularis*. Amer. Jour. Bot. 58: 292-299.
SMITH, G. L. and A. M. NOLDECKE.
  1960.   A statistical report on *A California Flora*. Leafl. West. Bot. 9: 117-132.
SMITH, R. H.
  1971.   Xylem monoterpenes of *Pinus ponderosa*, *P. washoensis*, and *P. jeffreyi* in the Warner Mountains of California. Madroño 21: 26-32.
SNYDER, L. A.
  1950.   Morphological variability and hybrid development in *Elymus glaucus*. Amer. Jour. Bot. 37: 628-636.

1951.   Cytology of inter-strain hybrids and the probable origin of variability in *Elymus glaucus.* Amer. Jour. Bot. 38: 195-202.

SODERSTROM, T. R. and H. F. DECKER.

1963.   *Swallenia*, a new name for the California genus *Ectosperma* (Gramineae). Madroño 17: 88.

SOLBRIG, O. T.

1965.   The California species of *Gutierrezia* (Compositae-Astereae). Madroño 18: 75-84.

SPELLENBERG, R.

1975.   Synthetic hybridization and taxonomy of western North American *Dichanthelium*, group *Lanuginosa* (Poaceae). Madroño 23: 134-153.

SPENCE, W.

1963.   A biosystematic study of the genus *Lessingia* Cham. Ph.D. Dissertation, Univ. California, Berkeley. vii + 145 pp. University Microfilms, Ann Arbor, Mich.

STANDLEY, P. C. and L. O. WILLIAMS.

1946-1976.   Flora of Guatemala. Fieldiana: Bot. 24.

STEBBINS, G. L.

1942a.   The genetic approach to problems of rare and endemic species. Madroño 6: 241-258.

1942b.   Polyploid complexes in relation to ecology and the history of floras. Amer. Nat. 76: 36-45.

1950.   Variation and Evolution in Plants. Columbia Univ. Press, New York and London. xx + 643 pp.

1952.   Aridity as a stimulus to plant evolution. Amer. Nat. 86: 33-44.

1953.   A new classification of the tribe Cichorieae, family Compositae. Madroño 12: 65-81.

1965.   Colonizing species of the native California flora. *In* Baker, H. G. and G. L. Stebbins (eds.), The Genetics of Colonizing Species, p. 173-195. Academic Press, New York and London.

1969.   The significance of hybridization for plant taxonomy and evolution. Taxon 18: 26-35.

1971.   Chromosomal Evolution in Higher Plants. Addison-Wesley, Reading, Mass. viii + 216 pp.

1974.   Flowering Plants. Evolution above the Species Level. The Belknap Press of Harvard University Press, Cambridge, Mass. xii + 399 pp.

STEBBINS, G. L. and A. DAY.

1967.   Cytogenetic evidence for long continued stability in the genus *Plantago*. Evolution 21: 409-428.

STEBBINS, G. L. and J. MAJOR.

1965.   Endemism and speciation in the California flora. Ecol. Monogr. 35: 1-35.

STERN, K. R.

1961.   Revision of *Dicentra* (Fumariaceae). Brittonia 13: 1-57.

STOCKWELL, P. and F. I. RIGHTER.

1946.   *Pinus*: the fertile species hybrid between knobcone and Monterey pines. Madroño 8: 157-160.

STONE, D. E.

1959.   A unique balanced breeding system in the vernal pool mouse-tails. Evolution 13: 151-174.

STRAW, R. M.

1955.   Hybridization, homogamy, and sympatric speciation. Evolution 9: 441-444.

1956.   Floral isolation in *Penstemon*. Amer. Nat. 90: 47-63.

1966.   A redefinition of *Penstemon* (Scrophulariaceae). Brittonia 18: 80-95.

STROTHER, J. L.

1974.   Taxonomy of *Tetradymia* (Compositae: Senecioneae). Brittonia 26: 177-202.

SUDWORTH, G. B.

1908.   Forest Trees of the Pacific Slope. Government Printing Office, Washington, D.C. 441 pp.

SWALLEN, J. R. and O. TOVAR.

1965.   The grass genus *Dissanthelium*. Phytologia 11: 361-376.

TÄCKHOLM, V.

1974.   Students' Flora of Egypt. Ed. 2. Cairo Univ., Cairo. 888 pp.

TADROS, T. M.

1957.   Evidence of the presence of an edapho-biotic factor in the problem of serpentine tolerance. Ecology 38: 14-23.

TAI, W. and R. K. VICKERY, JR.

1970.   Cytogenetic relationships of key diploid members of the *Mimulus glabratus* complex (Scrophulariaceae). Evolution 24: 670-679.

TALBOT, M. W., H. H. BISWELL, and A. L. HORMAY.
 1939. Fluctuations in the annual vegetation of California. Ecology 20: 394-402.
TAYLOR, D. W.
 1976. Disjunction of Great Basin plants in the northern Sierra Nevada. Madroño 23: 301-310.
THARP, B. C. and M. C. JOHNSTON.
 1961. Recharacterization of *Dichondra* (Convolvulaceae) and a revision of the North American species. Brittonia 13: 346-360.
THEIN, L. B.
 1969. Chromosomal translocations in *Gayophytum* (Onagraceae). Evolution 23: 456-465.
THEOBALD, W. L.
 1966. The *Lomatium dasycarpum—mohavense—foeniculaceum* complex. Brittonia 18: 1-18.
THOMAS, J. H.
 1961. Flora of the Santa Cruz Mountains of California. Stanford Univ. Press, Stanford, Calif. 434 pp.
THOMPSON, H. J.
 1953. The biosystematics of *Dodecatheon*. Contr. Dudley Herb. 4: 73-154.
 1960. A genetic approach to the taxonomy of *Mentzelia lindleyi* and *M. crocea* (Loasaceae). Brittonia 12: 81-93.
THOMPSON, H. J. and W. R. ERNST.
 1967. Floral biology and systematics of *Eucnide* (Loasaceae). Jour. Arnold Arb. 48: 56-76.
THOMPSON, H. J. and J. ROBERTS.
 1974. Loasaceae. *In* P. A. Munz, A Flora of Southern California, p. 549-559. Univ. California Press, Berkeley, Los Angeles, London.
THORNE, R. F.
 1969. The California Islands. Ann. Missouri Bot. Gard. 56: 391-408.
TILLETT, S. S.
 1967. The maritime species of *Abronia* (Nyctaginaceae). Brittonia 19: 299-327.
TOMB, S.
 1974. Chromosome numbers and generic relationships in subtribe Stephanomeriinae (Compositae: Cichorieae). Brittonia 26: 203-216.
TORRES, A. M.
 1964. Revision of *Sanvitalia* (Compositae-Heliantheae). Brittonia 16: 417-433.
TRAUB, H. P.
 1963. The genera of Amaryllidaceae. American Plant Life Society, La Jolla, Calif. 85 pp.
TUCKER, J. M.
 1952. Taxonomic interrelationships in the *Quercus dumosa* complex. Madroño 11: 234-251.
 1953. The relationships between *Quercus dumosa* and *Quercus turbinella*. Madroño 12: 49-60.
 1968. Identity of the oak tree at Live Oak Tanks, Joshua Tree National Monument, California. Madroño 19: 256-266.
TUCKER, J. M. and J. D. SAUER.
 1958. Aberrant *Amaranthus* populations of the Sacramento-San Joaquin delta, California. Madroño 14: 252-261.
TYRL, R. J.
 1975. Origin and distribution of polyploid *Achillea* (Compositae) in western North America. Brittonia 27: 187-196.
VAN HORN, G. S.
 1973. The taxonomic status of *Pentachaeta* and *Chaetopappa* with a revision of *Pentachaeta*. Univ. Calif. Publ. Bot. 65: 1-41, pl 1-5.
VASEK, F. C.
 1958. The relationship of *Clarkia exilis* to *Clarkia unguiculata*. Amer. Jour. Bot. 150-162.
 1964. The evolution of *Clarkia unguiculata* derivatives adapted to relatively xeric environments. Evolution 18: 26-42.
 1971. Variation in marginal populations oᶜ *Clarkia*. Ecology 53: 1046-1051.
VASEK, F. C. and R. H. SAUER.
 1971. Seasonal progression of flowering in *Clarkia*. Ecology 52: 1038-1045.

VICKERY, R. J., JR.
  1959.   Barriers to gene exchange within *Mimulus guttatus* (Scrophulariaceae). Evolution 13: 300-310.
  1967.   Experimental hybridizations in the genus *Mimulus* (Scrophulariaceae). V. Barriers to gene exchange between section *Simiolus* and the other main sections of the genus. Proc. Utah. Acad. 44: 316-320.
  1969.   Crossing barriers in *Mimulus*. Japan Jour. Genet. 44, suppl. 1: 325-336.
  1974.   Crossing barriers in the yellow monkey flowers of the genus *Mimulus* (Scrophulariaceae). Genetics Lectures 3: 33-82. Oregon State Univ. Press, Corvallis.

WAGENITZ, G.
  1969.   Abgrenzung und Gliederung der Gattung *Filago* L. s. 1. (Compositae-Inuleae). Wildenowia 5: 395-444.

WALKER, R. B.
  1954.   The ecology of serpentine soils. II. Factors affecting plant growth on serpentine soil. Ecology 35: 259-266.

WEBB, A. and S. CARLQUIST.
  1964.   Leaf anatomy as an indicator of *Salvia apiana-mellifera* introgression. Aliso 5: 437-449.

WEBER, W. A.
  1965.   Plant geography in the southern Rocky Mountains. *In* Wright, H. E., Jr., and D. G. Frey (eds.), The Quaternary of the United States, p. 453-468. Princeton Univ. Press, Princeton, N.J.

WEBSTER, G. L.
  1967.   The genera of Euphorbiaceae in the southeastern United States. Jour. Arnold Arb. 48: 303-430.

WEIL, J. and R. W. ALLARD.
  1965.   The mating system and genetic variability in natural populations of *Collinsia heterophylla*. Evolution 18: 515-525.

WEISSMAN, D. G. and D. C. RENTZ.
  1976.   Zoogeography of the grasshoppers and their relatives (Orthoptera) on the California Channel Islands. Jour. Biogeogr. 3: 105-114.

WELLS, P. V.
  1968.   New taxa, combinations, and chromosome numbers in *Arctostaphylos* (Ericaceae). Madroño 19: 193-224.
  1969.   The relation between mode of reproduction and extent of speciation in woody genera of the California chaparral. Evolution 23: 264-267.

WENT, F. W.
  1948.   Some parallels between desert and alpine flora in California. Madroño 9: 241-249.

WHALEN, M.
  1977.   Taxonomy of *Bebbia* (Compositae: Heliantheae). Madroño 24: 112-123.

WHITTAKER, R. H.
  1954a. The ecology of serpentine soils. I. Introduction. Ecology 35: 258-259.
  1954b. The ecology of serpentine soils. IV. The vegetational response to serpentine soils. Ecology 35: 275-288.
  1960.   Vegetation of the Siskiyou Mountains, Oregon and California. Ecol. Monogr. 30: 279-338.

WIENS, D.
  1964.   Revision of the acatophyllous species of *Phoradendron*. Brittonia 16: 11-54.

WIGGINS, I. L.
  1936.   A resurrection and revision of the genus *Iliamna* Greene. Contr. Dudley Herb. 1: 213-229, pl. 20.

WIGGINS, I. L. and D. M. PORTER.
  1971.   Flora of the Galápagos Islands. Stanford Univ. Press, Stanford, Calif. xx + 998 pp.

WILKEN, D. H.
  1975.   A systematic study of *Hulsea* (Asteraceae). Brittonia 27: 228-244.
  1977.   A new subspecies of *Hulsea vestita* (Asteraceae). Madroño 24: 48-55.

WILLIAMS, D. L.
1975. *Piptochaetium* (Gramineae) and associated taxa: evidence for the Tertiary migration of plants between North and South America. Jour. Biogeogr. 2: 75-85.

WILSON, R. C.
1970. The rediscovery of *Abronia alpina*, a rare specialized endemic of sandy meadows in the southern Sierra Nevada, California. Aliso 7: 201-205.
1972. *Abronia*: I. Distribution, ecology and habit of nine species of *Abronia* found in California. Aliso 7: 421-437.

WOLF, C. B.
1948. Taxonomic and distributional studies of the New World cypresses. Aliso 1: 1-250.

WOLFE, J. A.
1975. Some aspects of plant geography of the Northern Hemisphere during the Late Cretaceous and Tertiary. Ann. Missouri Bot. Gard. 62: 264-279.

WOOD, C. E., JR.
1961. A study of hybridization in *Downingia* (Campanulaceae). Jour. Arnold Arb. 42: 219-262.
1972. Morphology and phytogeography: The classical approach to the study of disjunctions. Ann. Missouri Bot. Gard. 59: 107-124.

WOODSON, R. E., JR.
1954. The North American species of *Asclepias*. Ann. Missouri Bot. Gard. 41: 1-211.

YOUNG, D. A.
1974. Comparative wood anatomy of *Malosma* and related genera (Anacardiaceae). Aliso 8: 133-146.
1974. Introgressive hybridization in two southern California species of *Rhus* (Anacardiaceae). Brittonia 26: 241-255.

YURTSEV, B. A.
1972. Phytogeography of northeastern Asia and the problem of Transberingian floristic interrelations. *In* A. Graham (ed.), Floristics and paleofloristics of Asia and Eastern North America, p. 19-54. Elsevier, Amsterdam, New York, London.

ZOBEL, B.
1951. The natural hybrid between Coulter and Jeffrey pines. Evolution 5: 405-413.

INDEX

117